Rough Framing Carpentry

by Mark Currie

Craftsman Book Company

6058 Corte del Cedro / P.O. Box 6500 / Carlsbad, CA 92018

Acknowledgments

The author thanks the following people, without whose support and good humor this book would be but a shadow of what it is:

Craig Campus	Jim Ford	Jeffrey T. May
Dana Currie	John & Chris Haddock	Richard McCluskey
Gene Currie	Craig Heyer	Gretchen Rymes
Ed De la Torre	John Jurkoic	Todd Schuster
Chris Dumont	John Krause	
Mark Flournoy	Chris Maher	Gunnar Waage

This book is dedicated to Roger Currie for being a friend as well as a father, and to Arliss Currie for giving me the perseverance to take things one step further.

The author and publisher thank *Stanley Tools,* a division of Stanley Works, New Britain, CT, for providing the framing square featured, and *Simpson Strong-Tie Company*, for providing the copyrighted illustrations of its products used in this book.

All photographs in this book were taken or conceived by the author, using a Canon F-1, a Hasselblad CM500, and a Pentax 6 x 7.

Thanks to Jennifer Dove Maurer for help with the cover.

Looking for other construction reference manuals?

Craftsman has the books to fill your needs. Call toll-free 1-800-829-8123 or write to Craftsman Book Company, P.O. Box 6500, Carlsbad, CA 92018 for a FREE CATALOG of books and videos.

Library of Congress Cataloging-in-Publication Data
Currie, Mark.
   Rough framing carpentry / by Mark Currie.
   p.   cm.
   Includes index.
   ISBN 0-934041-86-5
   1. House framing.   I. Title.
TH2301.C87   1993
694'.2--dc20   93-31738
                                     CIP

CONTENTS

Custom Homes, Tracts, or Commercial Work?

So you want to be a rough carpenter. Framers make good money doing healthy outdoor work. It's a great trade. Right?

That's right — with a few qualifications, of course. There *is* good money to be made, and the work is certainly done outdoors. But there's framing and then there's framing. Deciding that you want to be a framer isn't enough. There's another decision to make. What type of rough carpentry framing do you want to handle?

There are three basic types of rough carpentry:

- custom work

- tract work

- commercial work

Each requires unique skills and offers unique rewards. Since you'll probably want to specialize in one type of work or another, let's take a look at each of the three categories. As we do, keep in mind your skills and what you like and don't like. With a little help, you should have no trouble finding the niche where you belong in this trade, the place where you can work most happily and most productively.

While banging nails is the heart of all carpentry, the three types of rough carpentry work are very different. For example, on a commercial project there's almost certainly a rigid chain of command with lots of rules, plenty of supervision and very specific job requirements. Work on custom home jobs is seldom as carefully controlled. Tract jobs are different still. The hours you keep at a tract project might get you fired

Figure 1-1 Custom homes offer you a chance to learn how all the steps in framing a house interrelate with one another.

from a commercial or custom home building job. There are exceptions, but these generalizations about the three types of rough carpentry will apply on most jobs and nearly everywhere in the U.S.

Let's take a closer look at each type of rough carpentry work and see what generalizations will apply. I'll start with custom home work.

Custom Homes

Custom homes are usually built for an owner who plans to live in the home when it's completed. Working on a custom crew is the best way for a beginner to get into framing. And it's usually a rewarding experience. Now don't get me wrong, I'm not saying there are no slave drivers out there running custom crews. But as a rule, if you get to work on time and carry your part of the load, it's one of the least stressful framing jobs.

Skilled custom framers can expect 40 hours of work a week. Take-home pay tends to be much more predictable. That's an advantage, especially if you have a family to feed and a mortgage payment to make. Steady work on custom home jobs keeps many rough carpenters happy. Few would drop their position on a good steady custom crew because of rumors of "three hundred bucks a day" at

a tract down the street. A lot of people need the security of 40 hours of work a week. There's nothing wrong with that. If that's what you want, custom framing is probably for you.

Another advantage of custom framing is the variety of work you'll do. In a single day you may handle several dozen different tasks — from snapping to studs to sheeting. After a few years of custom work, you're bound to be a very well-rounded carpenter. Someone who spent those same years on a tract might be faster at specific parts of the job, but they won't have your understanding of how a complex building frame is assembled.

Figure 1-1 shows an elaborate custom house in the framing stages. With 14- to 20- foot 2 x 8 walls, this is the top end of the custom home market. At this point, the roof is being stacked, sheathing is going up on the outside walls, and arches are being constructed for the window openings. This house kept seven framers busy for about six weeks, at an average pay of around $15 per hour.

Custom home builders are notorious for requiring countless changes after construction has begun. That's usually due either to the owner's inexperience and lack of foresight, or a general contractor who didn't spell out exactly how much (or how little) house he was proposing to build. After all, it takes years of experience to visualize a finished home simply by examining the plans. Once the house begins to take shape, it's common for an owner to request alterations to make the house more livable. Few owners can appreciate the physical difficulty — and the added cost — of deviating from the plans. I've rarely built a custom home where changes weren't a problem. Dickering over changes and the cost of changes usually continues until the framing contractor receives his last payment from his final billing.

■ Spec Homes

There's a sub-category of custom home building that needs mention. Speculative (or *spec*) homes are built to be sold by the owner, not for occupancy by the owner. The developer is betting that the finished home can be sold for more than the cost of construction (plus the land). Compared to custom homes, spec houses are usually easier and less stressful work for the framer. This is because a spec often has fewer changes than a custom home. Investors build spec houses to make money, not to satisfy their own vision of a dream home. There's less romance involved. That's why changes aren't so common in a spec home project. The everyday decisions are usually made in a more logical (perhaps even a more sane) atmosphere.

If you're considering a career in rough carpentry, starting out on a custom home framing crew is a wise move. Sure, you'll spend the first few weeks just humping lumber from the lumber pile to the carpenters. We all did. But before long, you'll get an opportunity to show what you know, or at least how well you can listen. The more you learn, the more confident you become. Make yourself indispensable and the boss will begin assigning you more demanding work and rewarding you with heavier paychecks.

It's common to see framers on custom home jobs making $15 an hour after only a year in the business. Be a fast learner, a good listener, show some dexterity with tools and you'll do as well. In 1993, apprentice carpenters in my area were starting at $6 to $8 an hour. Tradesmen who supplied their own tools made more, of course. Top wage for a lead carpenter ranged from $18 to $30 an hour. Lead men are almost always required to supply tools.

Tract Projects

The design and construction of a housing tract is completely different from a custom home job. On most custom home jobs the goal is to put as much house as the owner can afford on as much land as the owner has. But when a developer designs a tract, it's a whole different ball game. The homes are usually intended to squeeze into a certain price bracket and to fit on the smallest lot the zoning ordinance will allow! Granted, some tracts offer true luxury homes at stratospheric prices, but most tract builders are very cost-conscious.

The whole thrust of tract design is to keep construction time and cost as low as possible. That's why most tracts have three or four basic designs, or models. All the models follow a consistent theme (Spanish-Mediterranean, French, contemporary, or ranch, for example). The architect makes small changes in the floor plan and exterior elevations so the fact that they're all pretty much the same isn't obvious. By adding a second story here, a window there, an arch somewhere else, and creating mirror images of each basic floor plan, the designer creates the illusion of 30 or more different plans. Look closer and you'll see there are only three or four basic plans.

There are advantages to framing the same home again and again. For one, practice makes perfect. The crew should get better and better as work progresses. Second, it allows specialization. That also increases productivity. The framing contractor may have ten or more separate carpentry crews, each crew of two or three carpenters handling a specific part of the framing process. It's almost like assembly-line production. As a crew finishes their part of the job, they move on to the next house and repeat that step again. One crew may frame all the walls, another may plumb and line the house. Then the joisting crew may take over, and on and on, until

Figure 1-2 When looking for work at a tract, check out what stage of framing they're in before you approach the foreman. This tract is half way through joisting, and just beginning to stack. They could probably use some extra help tipping up roof trusses.

the house is complete. After doing a few houses, each crew may be so familiar with their particular task that they could do it in their sleep. They can certainly do it faster, and thus, more cheaply.

The wood frame itself may be very similar on custom and tract homes. The difference is in the process. On a custom job, one crew does all the steps. On a tract project, there may be a different crew for each step. A tract builder needs specialists, not generalists.

Figure 1-2 is a typical tract in various stages of framing. The units in the foreground have just been plumbed and lined. Notice all the diagonal braces throughout the house. They are now ready for joisting of the second floor. The units in the right background have been joisted and the second story walls

(or *tops*) have been framed. The single story units in the background to the left have had their roof trusses rolled and are in the final stages of being sheeted. By studying the different degrees of completion in the tract units, you can tell which direction the crews are working and what step they're on (and might need help with). If I were looking for a job on this tract, I'd check with the floor joisting or roof stacking crews.

To minimize costs on tract work and to make the finished cost more predictable, the builder probably hires carpenters on a piecework basis. This arrangement gives the individual carpenters a direct incentive for high productivity. The more work they complete, the larger their paycheck is. Even if the tradesmen don't see a foreman all day long, they're sure to keep working hard.

Under a piece framing contract, you agree to do certain work at a set price. This price is usually based on the square footage of the house and the materials you're working with. For example, assume the contract price to build the walls in a 2000 square foot house is 30 cents a square foot. You'll get $600 for framing walls in each house, no matter how long the framing takes.

I've suggested that you'll become a more well-rounded carpenter on a custom framing crew. But on a tract job you'll learn more time-saving tricks in one week than you can in months on a custom home job. You have to — or you'll have the crew that follows yours breathing down your neck.

It doesn't take a rocket scientist to see why shortcuts and tricks are essential under piece framing contracts. If you can finish in six hours what once took eight, you just found two extra hours in a world short on time.

Of course, there are drawbacks. Piecers are renowned for leaving behind shoddy or unfinished work, hoping that the guy behind them will cover up their mess before the foreman finds it. There's a good chance the problems will come to light later and will have to be fixed. A large tract will have a whole crew dedicated to doing nothing but *pick-up*, or fixing unfinished or shoddy work before the house is inspected. But it's just as likely that most problems will escape detection (at least for a few years) when they're covered with stucco or drywall.

Every carpenter has had the experience of walking through a finished tract house and snickering at the bends and bulges in the drywall from the shoddy framing underneath. Tracts go up so fast that some of these problems are just to be expected. Defects that would interfere with the sale of a custom house are accepted as normal fare for most tract houses.

Figure 1-3 Commercial framers usually work large projects, such as this shopping center.

I've emphasized the differences between custom and tract framing. Which is right for you? Mostly it's a matter of attitude. I know framers who can handle both and actually prefer to alternate between the two types of work. If you're tired of having to build something over and over in what you consider an inferior way simply because your boss on custom home jobs grew up doing it that way, spend a few months on tract jobs doing it any way you want. Is the monotony of the tract driving you nutty? Jump to a custom crew for a while.

In the long run, you'll average about the same wages from either situation. While you might make more per hour from tract work, this is usually offset by the time waiting for more slabs to cure so you can begin work.

Commercial Work

Most commercial work is done on larger projects like shopping centers, malls, banks, schools, military housing, and multi-unit apartment, office or retail buildings. See Figure 1-3. Some of these

Figure 1-4   The building in the left background is being sheeted by one crew. When it's finished, the crew snapping, plating and detailing the building in the foreground will jump on it. At the same time, the wall builders now finishing up the building in the right background will start framing on the foreground building. If you don't keep a good pace going when you're piecing, you're going to get sandwiched fast.

projects use very little lumber. On others, several acres may be needed just to store the stacks of lumber delivered from the mill.

A lot of commercial work is done by framing contractors who have union contracts and use only union carpenters. If you want to work on those jobs, you'll have to join the local union. Only you can decide if that's best for you. It might be a good idea if union shops have most of the work in your area, especially when the construction industry is slow.

Probably the biggest difference between commercial carpentry and carpentry on custom or tract jobs is the size of the carpentry contracting company.

Most commercial framing companies are huge. For every person who considers this an advantage, there's another who would reckon it a curse.

Working for a company with two hundred carpenters on the payroll is very different from working for a company with ten or twelve. Yet some carpenters prefer to work for larger companies. Usually supervision is much more intense. Is that an advantage or a disadvantage? There are people with years of experience who prefer to have constant supervision. Some people will get absolutely nothing done if they're not supervised throughout the day. The world truly is divided into those who lead and those who follow. It's been my experience that commercial work, on the average, employs more followers than leaders.

A commercial job will usually have a crew for each step of the framing process. A foreman directs each crew and a lead foreman coordinates the work of all crews. Each crew has a few laborers who just move wood, get nails, help lift walls, and clean up. This is the ideal setup. When it's working well, the work gets done quickly and competently. Yet too many times things stray far from this ideal. A burnt-out foreman with a bad disposition (and there are many) can infect a whole crew with the same attitude. That breeds substandard performance.

When problems occur, no one is anxious to accept the blame. Bad feeling usually results. Also, conflicts between the trades are common on commercial jobs. When you have a dozen trades working in one area, you're going to have problems almost every day. That's rare on a custom home building site.

Most of the large commercial building sites I've worked on could be best described as chaotic. Doing piecework framing on a job like this can be a real test of patience.

Yet when the job's run right, commercial piece framing can be a gold mine. When a big job is running like clockwork, being part of it can be a treat.

Along with the disadvantages go some important advantages. Most large carpentry contracting companies offer good fringe benefits. It's common to see carpentry foremen on larger jobs driving late-model company-owned trucks with insurance and gas paid by the company. Even journeyman carpenters may receive pay for driving time and a gas allowance. These carpenters are probably enrolled in company-paid medical and dental plans and get paid holidays and vacations. Performance bonuses at the end of each job are common.

If you like the security of guaranteed work at one location for a long time, and need the benefits that small contractors can't supply, then commercial work might be for you. A major commercial project can keep you busy for months. For example, Figure 1-4 shows three buildings in a large condominium complex. These three stages will be repeated many times before the job is complete and may keep carpentry crews busy for an entire season.

Obviously there's a lot more to the rough framing business than I've covered in this chapter. But what I've explained so far should help you reach an important decision on an essential question. What type of rough carpentry is best for you?

Once you understand the categories of work, it's time to begin thinking about the tools you'll need. And that's the subject of the next chapter.

TWO

The Tools You'll Need

I've found that this is one of the hardest lessons to teach to novice carpenters: To make good money in this trade, you need good tools. And good tools aren't cheap. That shouldn't be news to you. But this may be: Good tools will help you make lots more money, lots easier and for longer.

When you're shopping for tools and equipment, buy the best you can afford. It hurts just once — at the cash register. Buy cheaper tools and it hurts again and again, every time a truly first-class tool would have done the job better, quicker and easier.

America may no longer be the world leader in all kinds of manufacturing, but we still make the best carpentry tools. That's why I recommend tools by the well-known U.S. suppliers. True, some imported tools cost less and you can often find them on sale. But I've seldom found imported tools that outlast comparable American-made tools. Obviously there are exceptions to this rule. If you can save a lot, it might be worth taking the chance. But consider carefully. And think about the lost productivity when a tool breaks. The morning you start on that $300-a-day job, you need reliable tools!

The Basic Hand Tools

To begin with, every carpenter needs a good tool and nail belt, usually called a *set of bags*. This is indispensable for carrying hand tools and nails. The bags worn by finish carpenters (and some rough framers) have

Figure 2-1 Here's a set of nail bags that are attached to the belt. With this type of belt there's no way to expand if you decide to carry more tools.

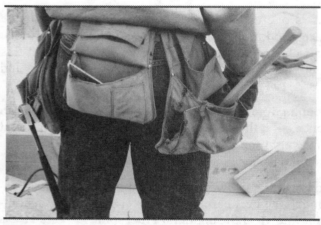

Figure 2-2 These are vinyl bags that are separate from the belt. If this carpenter decided to carry more tools or have more pockets for a wider selection of nails, he could simply slide another bag on to his belt.

pouches permanently attached to the belt (Figure 2-1). But most rough framers prefer separate leather bags that aren't attached to the belt. Although the separate bags cost more, they let you carry more tools — and carry them more comfortably. A tool belt with up to four detachable bags can carry different sizes of nails and the most frequently-used hand tools. Most framers carry two or three bags and a hammer holder on their tool belt.

Fashion-minded framers who can't do their best when they don't look their best will be relieved to know that several companies now sell vinyl bags (Figure 2-2). These come in a variety of colors, from earthy brown to Day-Glo orange, to harmonize or contrast with any type of boom box.

It's a temptation to carry too many tools in your belt. I'm amazed at what some of the carpenters on my payroll carry around all day: planers, 6-inch levels, Grateful Dead tapes. One fellow I know even carries two hammers. Granted there are days when you need a planer and a 6-inch level. But why carry a second hammer, a level and a planer all day? Better to leave in the truck any equipment that you use less than daily.

Here's my list of the bare minimum hand tools to carry:

- a 25- or 30-foot tape

- a chalk line (I like one that's geared to reel in quickly)

- a square (preferably a speed square)

- a nail puller or "cat's paw"

- pliers with nail cutters at the base of the blades

- a utility knife

- a chisel you're willing to use on nails and concrete

- a keel and a pencil

- a good framing hammer or axe

Every carpenter has personal preferences on tools. Take hammers, for example. On a larger job you'll probably find someone using an axe, someone using a 20-ounce hammer, a 24-ounce, a 28-ounce, and even the caveman special, the 32-ounce club. Some hang their hammers from the sides of their bags and get a surprise wallop to the knee now and then. Others hang a hammer right above their rear ends. Some stuff the hammer headfirst into their bag so the hammer can drop out on a buddy when they're climbing a wall.

Figure 2-3   A beam saw will allow you to make deeper cuts than an ordinary circular saw. Always keep the pressure forward, and watch that unguarded blade!

the worm-drive Skil™ saw even though it's nearly twice as much money as sidewinder types. My experience is that these sidewinder saws aren't made to take hard daily use. They tend to slow me down and that's expensive.

A lot of carpenters modify the saw by replacing the 6-foot cord with a 50- to 100-foot cord. With the short cord, you're constantly unplugging and replugging your saw as you move around the work space. The long cord saves time. But it may also void the warranty on a new saw. So beware. A worm-drive saw will cost about $120. With a long cord and a few extra blades, expect to pay about $150.

There's one more tool I didn't mention. It's not on my list above but I hope you find it essential for framing. I always carry a small calculator. I use a cheap, solar-powered model that sells for about $15. With this little gem I can figure rafter cuts as fast as anyone using one of the $300 idiot-proof models. I view a calculator as a valuable tool rather than a crutch. A carpenter who can use a calculator (and always has it on him) is worth more to me than someone with equal experience who doesn't use one.

A set of bags with all your basic hand tools is going to cost between $100 and $150.

The Power Tools

Your most useful power tool is bound to be a circular saw. Many types are available. But most of the carpenters I know use a worm-drive Skil™ saw and prefer that brand over the competition. Makita offers a lighter and cheaper saw that's similar and is popular with some carpenters. I can recommend

When you're using a circular saw, here's the most important rule: never ever move the saw backwards. Always use a forward motion. The split second you let your saw move backwards, it'll be in your lap. If you're new to a circular saw, keep this in mind on every cut. The saw is built with lots of torque and will jump backwards before it binds.

Another commonly used saw is the *beam saw*. This is nothing but an oversized circular saw. It's large blade allows you to make deeper cuts than your typical 7-inch circular saw. The most common beam saw is made by Makita (Figure 2-3). Beam saws will cost from $250 to $600, depending on the size and the quality of the blade you choose.

Most contractors supply you with a beam saw if you're doing much beam work. Although beam saws may look a bit frightening at first, with their imposing blades and overall size, they're actually very easy to use. Just make sure your material is secure before you cut, and keep your body parts away from the unguarded blade as it passes beneath the beam. A common circular saw will jump back

Figure 2-4   For moving headers, a reciprocating saw is indispensable.

The Nice-to-Have Tools

Most framing companies want their employees to show up with the tools I've described — the hand tools and circular saw. You supply those and your employer will supply any other tools necessary to complete the project.

Even so, the longer you're in this trade, the more tools you're likely to acquire. With more specialized tools, you're prepared to handle more specialized work at higher production rates. Among the first "extra" tools you'll need will be a reciprocating saw (commonly called a Sawzall), drill, ramset, ladder, compressor and nail gun. Most employers will pay extra for the use of tools like these on a job.

A *reciprocating saw* has a blade that reciprocates, rather than spins. They're a must in some situations where a circular saw just won't do the job. For example, one common job for a framer is to lower a door header. With a Sawzall you can insert the blade between the king stud and header and cut all the nails (Figure 2-4). Without one, you'd be struggling with a nail puller. Also, cutting out the bottom plate of a door opening that's in place can only be done with a Sawzall (Figure 2-5). (Actually the plate could be cut with a circular saw before you raise the wall. That's what I recommend. The Sawzall method takes much longer.)

If you do tract work, I recommend investing in a compressor and nail gun (Figure 2-6). Nothing increases productivity as much as a nail gun. It's like adding one employee to the crew because it just about doubles the work one carpenter can produce. Brand new, a top-of-the-line compressor, gun and hose will set you back $1200. Emglo makes a very popular and reliable line of electric and gas compressors. The only framing nail gun I'd ever buy is the Hitachi. It's lighter, better balanced and cheaper or comparably priced to other guns. Used equipment is always available and I'd encourage going this route if you can't afford or finance a new rig.

at you when it binds — that's how most fingers get severed. A beam saw is more predictable. If it binds, it usually just slows down or stops.

So now let's stand back and take stock. For about $250 you're equipped with all the tools necessary to look like a real pro — even if you haven't a clue about carpentry work!

Don't go looking for work as a carpenter with a bag full of funky tools. Good tools help identify a good craftsman. They communicate professionalism in the carpentry trade the same way a smart business suit communicates professionalism in the business world.

Figure 2-5   A reciprocating saw can make cuts that might be impossible with a circular saw.

Some carpenters aren't willing to invest in a compressor, hose and nail gun. Yet these same tradesmen will spend $800 a month on a new truck. If you're interested in making as much money as you can, get a nail gun. Talk to dealers about financing. Put it on your credit card. Inspire your relatives to loan you half, if necessary. You'll be doing yourself a big favor!

It takes time and money to acquire a good collection of tools, and most companies understand this. I know many carpenters who have made a good living working both tracts and custom home jobs while never owning more than the basic hand tools and a circular saw. But if you have a secret desire to be a licensed contractor some day, start tooling up as soon as you can. Then, when you start bidding jobs on your own, the extra investment will be minimal.

Figure 2-6   For the serious framer, a compressor and nail gun are mandatory. You could probably pick up an old workhorse like this for $300 or $400 complete.

THREE

Hardware and Materials

Take a few minutes to stroll through the hardware department the next time you visit a large building material retailer. You may be astonished at the almost endless variety of framing hardware we carpenters are supposed to use: straps, connectors, clips, brackets, braces, plates, hangers, anchors, ties and angles. And what you see on the shelves is just the tip of the iceberg. Most larger retailers have catalogs full of the kind of framing hardware that's required on some blueprints. A lot of it's available only on special order.

Identifying the Hardware

When you're starting out, you'll probably feel a little intimidated by all this hardware. How are we supposed to use this stuff? And which size and finish are we supposed to be using? Worse, what's used and how each contractor installs it varies from one area to the next, from one state or city to another. If you move from Arizona to Florida, or New Jersey to Oregon, the types and amounts of hardware you install is going to change completely. Even the hardware that looks familiar probably goes by a different name. And there are bound to be a few items you've never seen before.

The framing hardware capital of the world has to be earthquake country — California and the entire Pacific Coast. The earthquakes that punish this part of the country provide engineers with abundant opportunities to experiment with our safety. Scientific data seems to

Figure 3-1 Here's a shear transfer that just won't quit! The floor joists are connected to the glu-lam with A35s at about 10 inches on center. Then straps are bolted to the glu-lams and up to the base of a 6 x 6 built into the wall. Then a strap is nailed to the top of the 6 x 6 and awaits another post in the second floor walls. The shear is nailed 3 inches on center around all the edges, and 12 inches on center into the field studs. The posts make up the rough opening of the windows.

show that there's a direct relationship between the weight of framing hardware in a wall and the probability that a wall will be left standing after a quake.

Whether it's fear of earthquakes or the fear of a lawsuit, engineers really have a ball in California. Massive straps are welded and bolted and nailed to buildings (Figure 3-1) like stripes on a tiger. Saddles connect 16-inch posts to ridges and glu-lams like braces on the door of a bank vault (Figure 3-2). Hips and ridges are strapped and restrapped (Figure 3-3). On every horizon you can see the shiny reflection of endless rows of A35s glistening in the cloudless sky. And by golly, if you mess up and leave one washer off of one

hold-down, the inspector is going to react as though you've exposed the city to liability that could bankrupt local government for generations.

While most framers in California feel that the engineers and inspectors are over-reacting to cover their personal and municipal liabilities, there are places on the West Coast where things work quite differently. I did some residential remodeling on a peninsula in the state of Washington and found virtually no hardware used at all. Until recently very little framing hardware was used in New Mexico. On common, light commercial framing jobs in New Mexico, builders use joist hangers, an occasional post cap, and not much else. With the custom

Figure 3-2   This 3800 square foot house has over $2000 worth of hardware in it. These posts hold up ridges with an ECC66, and are connected to glu-lam beams with another CC66 turned upside down.

houses, such as you'd find in Santa Fe, most of the structural members are oversized exposed beams or round vigas (unmilled 6-to 12-inch trees with the bark removed). Your typical steel post cap would be unacceptable from an aesthetic standpoint. Beams are connected to one another using combinations of "Z" lap joints, mortise, and tenons (Figure 3-4). This system is sometimes referred to as Japanese joinery. Beams are connected to posts with vertically-run 16-inch lag bolts. The end product is more of a "crafted" than "framed" house.

Common Framing Hardware

Although there's a large selection of hardware available, a typical job will only use about ten different types. Figure 3-5 shows several of them. Hardware is used to strengthen load areas. Designers use large beams to create open, unobstructed spaces in a house. Each beam concentrates a large amount of weight on two or three posts, spreading the weight of the wall it's supporting along its entire length. When there's so much riding on only two or three posts, these supports become critical. The designer will specify hardware to connect the beam to the posts.

Hold-downs

The plans may also call for hardware to connect the post to the slab. This is typically done with either a PA, HPAHD, or a HD strap. A *PA strap* is simply a steel strap that's partly submerged in the concrete. It's lighter duty than the others. Once the wall is standing, the post installed and any wall sheeting applied, you nail the strap to the post with 16d nails. Sometimes shear walls will have a PA strap at either

Figure 3-3   Here is another lawsuit-paranoid engineer's dream. An EPC66 holds the post to the ridge. Then a MST60 holds it again, this time lapping over the ridge. Then a combination of two more MST60s and an ST48 hold the hips to the ridge. Each ridge was detailed with this amount of hardware.

Figure 3-4   This "Z" lap joint in the beam is an example of Japanese joinery.

end to hold the entire wall in place in case of an earthquake. An HPAHD strap is simply a PA strap that's bolted instead of nailed to the post.

A hold-down (HD) is used in the same places as a PA strap, but there are three main differences between them. HDs are stronger, more expensive, and a lot more time-consuming to install. That's because they're bolted to the post, instead of being nailed. And that takes longer. The way that HDs are designed, there is only a small margin for error. The base of the HD has a hole for the anchor bolt, which attaches it to the slab. In most cases, the post that the HD is attached to must be set in a specific spot, with only a little play. If the concrete person doesn't understand that the stud bolt must be placed in an exact location, you could face big problems down the line.

On an organized job, the lead supervisor will coordinate with the framer on locating all hold-downs before the concrete is poured. The last job

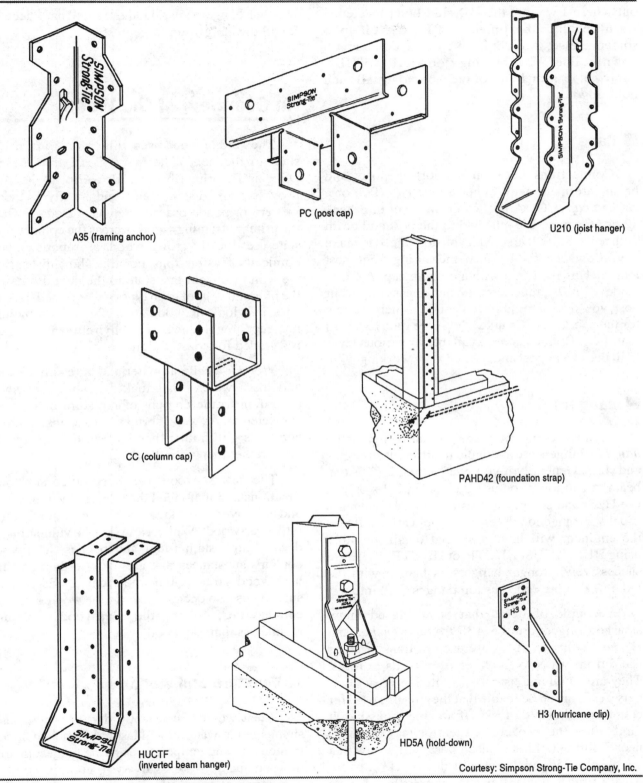

A35 (framing anchor)

PC (post cap)

U210 (joist hanger)

CC (column cap)

PAHD42 (foundation strap)

HUCTF
(inverted beam hanger)

HD5A (hold-down)

H3 (hurricane clip)

Courtesy: Simpson Strong-Tie Company, Inc.

Figure 3-5   An assortment of typical framing hardware.

I framed had 40 HDs. It took me six hours to mark out where I wanted the HD stud bolts located. I wasn't paid for this time, yet if I had left it up to someone else, it might have cost me 60 hours on down the line! When bidding a job with lots of HDs, I allow at least an hour for the labor to install each hold-down.

■ Caps

Connections between supporting posts and beams are made with either a *post cap* (PC) or a *column cap* (CC). A post cap is made of thin steel, 12 or 16 gauge. It's attached to the post and beam with nails. Since it is so thin, you don't need to make any allowance for it when measuring your post length. This isn't true with a column cap. A CC is made with 7 gauge steel. When measuring your post, you'll need to subtract a good 3/16 inch to allow for the cap. And CCs take a lot more time to install than PCs. Both caps are available in various sizes to fit the size of post and beam you're working with.

■ Hangers

Another common piece of hardware is the *hanger*. Hangers are available in different strengths and sizes. Generally if you're attaching 2 x 10s to a beam or rim joist, you use U210s. With 2 x 6s you use U26s and so on. The designated "U" hanger is usually strong enough for most applications. If not, the engineer will have a special detail, possibly using HUS, HHUS, HUCTF, or HHUTF. These are successively stronger hangers. A hanger with the letters TF in the code has top flanges built into it.

A couple of items that have caused many smashed fingers are the A35 *framing anchor* and H3 *hurricane tie*. A35s are used to transfer shear force from walls to floors or from walls to roofs. They are small and usually installed in great numbers. A35s are so bendable that they're also used for countless other little tasks. If an area looks like it might draw the inspector's attention (say the engineer didn't detail in at least a post cap), throw in a couple of A35s and that usually does it. H3 clips are usually used to hold rafters or trusses to the top plate

of a wall. Both of these are attached with 1½- or 1¼-inch hanger nails, and these small nails account for all the banged-up fingers.

An Overview of the Wood

The kind of wood used will also vary from one state to the next. The West Coast frames with Douglas fir primarily, and also a lot of hem-fir. Move a short distance east and locally-abundant western or ponderosa pine replaces Douglas fir as the primary framing wood. Across the country each state uses locally-grown wood, or imports an economically feasible one. For this short discussion, we won't go from one state to the next discussing the pros and cons of each type of wood used. Instead let's just look at wood in general; its structural makeup, how it reacts to its environment, and how it's graded for construction use.

Wood is a cellular material that readily absorbs and loses water. That makes it shrink or swell according to the amount of moisture or heat it's subjected to. As wood shrinks or expands, it moves across its girth rather than its length — across the grain rather than with the grain.

The newest wood in a tree, called sapwood, functions mainly to feed the tree. As the tree grows and new layers are added, the sapwood slowly turns into heartwood. While sapwood is vulnerable to decay fungi, stain fungi and beetles, heartwood contains substances that are toxic to them. So the heartwood protects the tree from disease. These substances also occupy space in the wood's interior cell cavities, making the heartwood swell and shrink less than sapwood.

■ Moisture and Wood

When wood is constantly dry or continuously submerged in water, it will last indefinitely. Change in moisture and temperature are the biggest factors in wood decay. Water is always present in wood — how much or how little depends on the humidity.

Once a tree is cut down and the wood milled, the cell cavities in the wood release their bound water, causing the wood to shrink. As the wood moves from one environment to another, it strives to keep in balance with the moisture in the atmosphere. It is important to point out that this is true even with kiln-dried wood. Wet wood subjected to a dry atmosphere will dry quickly, and dry wood (even kiln dried) in a wet (or humid) atmosphere will regain moisture.

Ideally, all framers should use wood that's dry and has been allowed to stabilize at the building site. But as a practical matter, most framers can't allow lumber to season properly before using it in the framing. It just takes too long.

During my training as a framer, no one ever advised me to check the amount of moisture in wood before writing out a check to the lumberyard. Of course I encountered the usual problems with wood that was too green: I made beautiful miter cuts only to watch them open up as the wood dried. Warping wood, dimensional changes, opened glue joints, raised grain, and end checks are all caused by too much or too little moisture during construction. All of this can be avoided if you check the moisture content of the wood before you buy it. This is done quickly and accurately with a small, hand-held meter.

This meter measures the electrical conductivity of the wood to give a percentage reading of the moisture content. Here's how it works: Wet wood conducts electricity, while dry wood is an electrical insulator. Depending on where you're working, air-dried wood becomes stable in the 12 to 15 percent range. Interior kiln-dried wood should be around 6 to 8 percent. This varies with geographical area. A board in the Northeast might stabilize at 14 percent. Take it to the Southwest, and it stabilizes at 6 percent.

Hand-held meters run from $100 on up. Many contractors buy tools that cost four times this much to give them cuts and joints that are later ruined as the wood dries and shrinks. If they knew the exact properties of the wood they were working with, they would be assured that their beautiful miter cuts would remain tight.

Quartersawn vs. Plainsawn

When a log is milled, the boards are either quartersawn or plainsawn. A *quartersawn* (or edge-grained) board has growth rings that form an angle to the surface of the board of 45 degrees to 90 degrees. In a *plainsawn* (or flat-grained) board, the angle is less than 45 degrees. A quartersawn board shrinks more in thickness and a plainsawn shrinks more in width. When a board shrinks in width, it tends to cup, twist and generally distort (Figure 3-6). Woodworkers value a quartersawn board because it's more stable dimensionally.

Generally rough framers don't care too much whether the lumber is quartersawn or plainsawn. Now and then you might run across a stud that's severely twisted. It's probably a plainsawn board. With exposed wood, such as the fascia board, this might be a problem. A severely cupped piece of fascia is only going to get worse after it's installed and dried by the sun. You also have to be careful with deck railing. If you have the advantage of handpicking your material from the lumberyard, always load up on the quartersawn wood!

Lumber Grades

When a tree is harvested, only about a quarter of it is used. Half is left behind in the forest, and half of what makes it to the mill is lost in trimming. As the individual pieces of lumber roll off the saw, they represent a wide range of strengths and appearances. Because the mill's main objective is to produce as many high-quality boards as possible, it might even rip a board in half and throw half away to create a higher-grade board out of the other half.

Originally, each sawmill sold its lumber locally and grades had only local significance. As lumber began being transported and competition between different mills heated up, this system just didn't work anymore. New standards were adopted in 1925. Since 1970, nearly all lumber used in construction is graded under the American Softwood Lumber Standard PS 20-70. Under these standards, all construction lumber is placed in one of three categories: stress graded, nonstress graded,

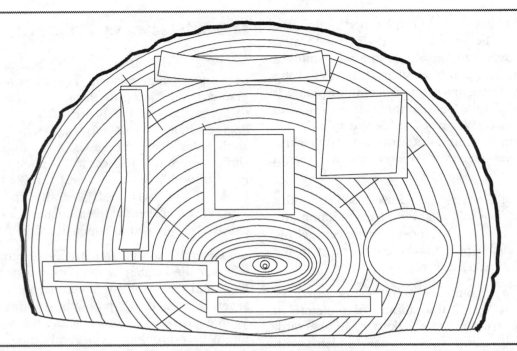

Figure 3-6   This shows how the tree's annual rings affect the shrinkage and distortion of flat, square and round cuts.

and appearance lumber. Stress graded and nonstress graded deal with the strength of the lumber. Appearance grade is used where the appearance of the wood is of primary importance.

For lumber classed as stress graded, a grader watches for factors that influence strength, including density and average weight, decay, slope of grain, size and location of knots, checks, and splits. With appearance lumber, graders assume that the board will be cut up into smaller pieces. A board is graded on how many imaginary pieces you can squeeze out of it, in between the knots and small defects. The higher the grade, the wider and longer the clear cuttings between the knots will be. It's not a matter of what shape the defects are in, but where they're located!

Stress-Graded Lumber. Dimension lumber is lumber that's at least 2 inches, but not more than 5 inches, in nominal thickness. It's the most common stress-graded lumber in use today. Dimension lumber is divided into five categories: Light Framing, Structural Framing, Studs, Structural Joists, and Appearance Framing.

Light Framing material usually refers to 2 x 4s. The grades available, from strongest to weakest, are *Construction, Standard*, and *Utility*. A grade marked "STD&BTR" means that originally the lumber was graded both Standard and Construction. In other words, for any given number of individual boards, a certain percentage will be Standard and the rest Construction grade.

Structural Framing material includes 2 x 6, 2 x 8 and 2 x 10, and for all purposes will include Structural Joists as well. They are graded *Select Structural, No. 1, No. 2*, and *No. 3*. The grades refer to differences in bending strengths as might be experienced in various rafter and joist applications. Again, a lot of structural lumber is graded in combinations such as No.2&BTR or No.3&BTR.

Stud Grade is most commonly 2 x 4s but can go on up to 4 x 4s. The stud grade can also be identified as *PET*, which means it was precision trimmed. That assures you each end was cut square and the board is the exact length you order.

Appearance Grade is based on the absence of knots and surface blemishes, not strength.

Nonstress-Graded Lumber. This category of lumber is sometimes referred to as *yard lumber*. For most of this century, most wood used by the construction industry wasn't stress graded. The most common nonstress-graded lumber is known as 1 x (one by). That's a nominal 1-inch thickness, or ¾ inch when both sides are dressed. It's available 2, 3, 4, 6, 8, 10, and 12 inches wide. Grades No. 1, No. 2, No. 3 (Construction, Standard, and Utility) are readily available. The lumber is prepared with square edges, tongue and grooved, or with a shiplapped joint.

Nonstress-graded lumber is also identified either as *Selects* or *Commons*. Selects, graded on appearance, are used for siding, overhangs, cornice work, or shelving. Commons have more knots and knotholes, and are used for subfloors, skip sheeting roofs, and let-in braces.

Appearance Lumber. Appearance lumber is also nonstress graded. It falls into a separate category because the emphasis is mainly on looks. Appearance grade includes most lumber that's milled with a pattern, such as baseboard, window casing, flooring, and paneling. Most appearance lumber is graded by letters *(B&BTR, C&BTR)*, and the descriptive terms *Prime* and *Clear*. You might also get the option of *FG* (flat grain or plainsawn), *VG* (vertical grain or quartersawn), or *MG* (mixed grain — a combination of both).

■ How Grading Applies to the Framer

With all this talk about lumber grades, it's easy to feel overwhelmed. But there are only a few instances where you really need to watch out. When in doubt about any practices, ask the local lumberyard salesmen. They order wood for framers all day long, so they know what's being used in the area.

When ordering wood for a job, first consider the floor joists and roof rafters. They're engineered for the live and dead weights they'll carry. The plans will call out the size and grade of wood to use. For instance, if you find the floor detail, it'll show the directions of the joists, and then call out *2 x 12 No.2&BTR 16" O.C.* You would order enough No. 2 or Better 2 x 12s to be laid at 16 inches on center. Other designations might be *2 x 12 S.S. 16" O.C.* (2 x 12 Select Structural) or 2-2 x 12 No.1&BTR 12" O.C. (two No. 1 or Better 2 x 12s at 12 inches on center). That last one can be a killer to miss. I worked for a company that didn't realize the first floor of a three-story building needed double 2 x 12s until the building was complete. There was this suspiciously tall pile of 2 x 12s left over. We went back and doubled up each joist, but I wouldn't want to have to do that again. Remember that you can always use a higher grade than is called out, but never a lower grade.

Lumber that you'll use for fascia, deck rail, or exposed ceilings is available two different ways, rough (resawn) or smooth (dressed). The plans should tell you the type to order. Rough lumber is usually thicker and wider than standard wood. A 4 x 4 might measure 4 by 4 inches, rather than the typical 3½ by 3½ inches of a dressed board. A resawn board is a board that has been dressed (made smooth), then sawn again to make it rough again. Order resawn wood with the sides and edges in mind. For instance RS1S2E is a board with one side and two edges resawn or rough. You'd probably use lumber like this for fascia. Because the backside is rarely seen, it's left smooth, with only the front side and the two edges rough. For smooth wood, you can order lumber that's surfaced (planed) S1S, S2S, S1E, etc. Most wood is milled S4S (smooth four sides). That makes it the most economical.

Both the hardware and materials required to build a house to code vary from one building department to the next and from state to state. Rely on local engineers and architects to familiarize you with regional customs and laws. The county building department can provide you with rafter and joist span tables, as well as hardware requirements. But the best way to learn about the hardware and materials used in this trade is to get out there and get your hands dirty using some of it.

FOUR

Snap, Plate and Detail

The first step — and one of the most important — in any framing job is snapping and detailing. In the old days of framing, the framers tackled each wall separately. They bolted a bottom plate (or sill plate) to the foundation, raised a stud on either end, located and installed any windows or doors, then filled in the remaining studs. This process, known as *stick framing*, is still common in some areas. But for the serious framer, it's so slow that the cost is prohibitive at today's wages.

Now we concentrate on each step in the process, not each wall. Before assembling the first wall, we mark out each wall on the floor, cut and temporarily nail together a top and bottom plate, and locate and detail each component of the wall. *Detail* means writing the locations on the top and bottom plates. This detailing is the brain work in wall building. It directs the carpenters who actually frame the walls. Once it's done, it just takes muscle and the ability to follow directions to build and raise the entire group of walls.

The first step, *snapping*, is locating and marking out each wall and its exact intersection with other walls. The snapper uses a chalk line to *snap out* lines that indicate the location of the interior and exterior walls on the concrete slab or wood floor. Any mistakes made here will show up again and again, from the floor tiles to the roof stacking. Walls snapped out of square or in the wrong location can be a disaster that you may not recognize until it's too late.

That's why the job of snapping and detailing is usually reserved for the most experienced carpenter on the crew. You can use a less-experienced person to do the more labor-intensive step of plating. On a custom crew, either the company owner or his most trusted lead man will probably be doing the snapping and detailing.

Figure 4-1   A clean slab that's ready to be snapped.

Figure 4-2   To construct a sguare building, snap two perpendicular walls using a 3, 4, 5 triangle.

When the entire house is plated, detail all the doors, windows, studs and special notes by transferring them from the plans directly onto the plates. The detailer spells out the exact location of everything that belongs in each wall, according to the plans provided. This sounds simple, but that's not to say it's easy!

The framing supervisor should do some planning before turning over the plans to the snap, plate and detail crew. Among the items detailed are door and window header and cripple sizes, medicine cabinets, layout for shear walls, siding or elevation backing, fire blocks, posts, hold-downs, and studs.

The supervisor should highlight these detail items on the plans. Anything left out during the detailing will be harder to add later. On a piecework job, you'll be back-charged if someone has to go back and install something you missed.

Snapping

If the house is built on a slab, begin by sweeping the slab clean of all debris. Figure 4-1 shows a slab ready for snapping. To start, you want to find the two longest exterior walls that are perpendicular to each other. In other words, they form an outside corner. You'll use them to create a perfect square.

■ Squaring the Walls

Nearly all walls are built at a right angle to some other wall. The trick is to create a perfect square out of two walls and then measure every other wall off of them to create a uniform layout. If you create the square with the longest walls possible, you can measure all the other walls off of them. Don't ever trust the concrete to be square! I've seen it up to 6 inches out in 20 feet. If you go by the existing concrete, be prepared to look like a fool when the tile man gets going.

The easiest way to construct a square corner is to use the Pythagorean Theorem: In a triangle with short sides A and B and a long side of C, if A squared plus B squared equals C squared, the intersection of the short sides is a perfect 90-degree corner. This is commonly called a 3, 4, 5 triangle because 9 (3 squared) plus 16 (4 squared) equals 25 (5 squared). Figure 4-2 shows a 3, 4, 5 triangle. You can multiply the numbers for the side lengths by 2 or 3 if you have that much room to work on the slab. For example, the side lengths can be 6, 8 and 10 or 9, 12 and 15. The longer the sides you measure, the more accurate the job will be.

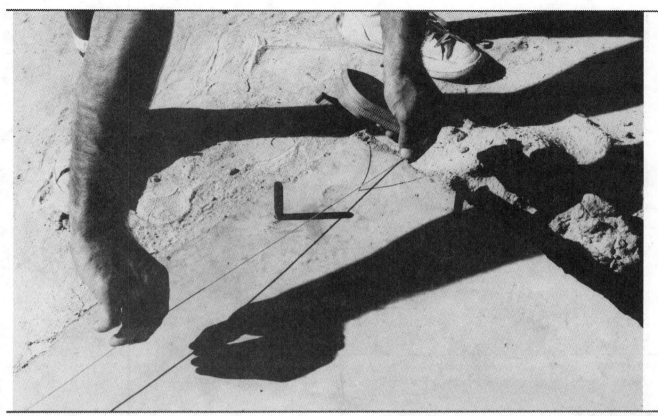

Figure 4-3　First snap the longest running exterior wall all the way through, 3½ inches in, from one corner of the slab to the other.

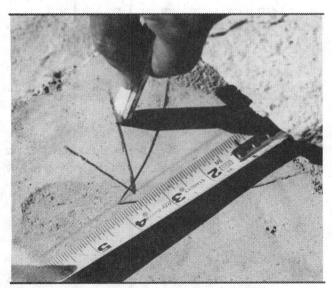

Figure 4-4　Then pull 3½ inches in, along the line you just snapped, and make a mark, This will be the inside corner of the room. From the inside corner pull an additional 3 feet along the snapped line, and make a mark.

For a 2 x 4 wall, measure in 3½ inches from each corner where the wall will stand. Snap a line between the two marks (Figure 4-3). If it's a 2 x 6 instead of a 2 x 4 wall, measure in 5½ inches. Then find the longest exterior wall that runs perpendicular to the wall you just snapped. Measure in 3½ inches (or 5½) from the edge of the concrete along the first wall you snapped and make a mark with your pencil (Figure 4-4). That marks the inside corner where these two walls will intersect.

I'll assume you're using a 3, 4, 5 triangle. Begin by pulling a tape 3 feet along the snapped line from the inside corner you created. Make a mark. Then pull your tape from the same corner but along the direction of the second wall (perpendicular to the first wall). Sight your tape so that it looks square off the first wall, and swing a mark at 4 feet (Figure 4-5). Make the mark long enough that you can find it when measuring the diagonal (Figure 4-6). Now

Figure 4-5   Now, holding your tape perpendicular to the 3-foot wall (or the first wall you snapped), pull from the inside corner 4 feet out in the direction of the second wall (perpendicular to the first wall). Make a mark.

hold your tape on the 3-foot mark and pull toward the 4-foot mark, creating the diagonal of the triangle, as in Figure 4-2. Clearly mark the exact point where the 5-foot measurement on your tape inter-

Figure 4-6   When you make the 4-foot mark, make an actual line by sweeping your tape back and forth an inch either way. This gives you plenty of room to find the intersection with the 5-foot mark on your tape.

sects the 4-foot line on the slab (Figure 4-7). That's the point where the wall line will intersect when you snap it from the inside corner. Go ahead and snap a line from the inside corner through that mark as far down the slab as you need (Figure 4-8).

■ Continuing the Snapping

Once you've created a perfectly square corner with these two walls, go with the measurements on the plans to locate the position of all the other walls relative to the first two. That ensures a uniform and square building — and a lasting friendship with your friend the tile man.

First, identify the stud size. Usually the studs are 2 x 4s. But don't take that for granted. Look for a special note next to the wall, or possibly a note elsewhere on the plans that might call out "All Exterior Walls 2 x 6." Many exterior walls in colder climates are made from 2 x 6, 2 x 8, or even 2 x 10 studs so the wall can enclose thicker insulation.

The plans will call out the measurements in feet and inches between the walls in each room. I like to snap out the long hall walls first. If you're working alone, drive a 1½-inch Teco nail (hanger nail) or short concrete nail into the concrete to hold the end of your chalk line (Figure 4-9). Use a lead weight to hold the chalk line if you don't have Teco nails or if the concrete is too hard to nail into.

When you're snapping interior walls, be sure to note which side of the wall you're measuring to and from. A wall marked on the wrong side of the line may make the hall too narrow to comply with the local building code. That's a costly mistake! Always double-check which side of the wall the plan is telling you to measure to, or from.

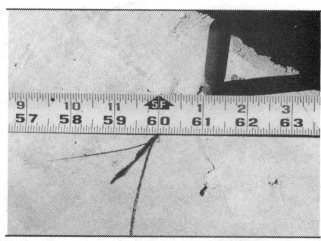

Figure 4-7   Lay your tape on the slab and stretch it from the 3-foot mark you made. Bring your tape over until the 5-foot mark on the tape intersects with the 4-foot line on the slab. Mark this intersection.

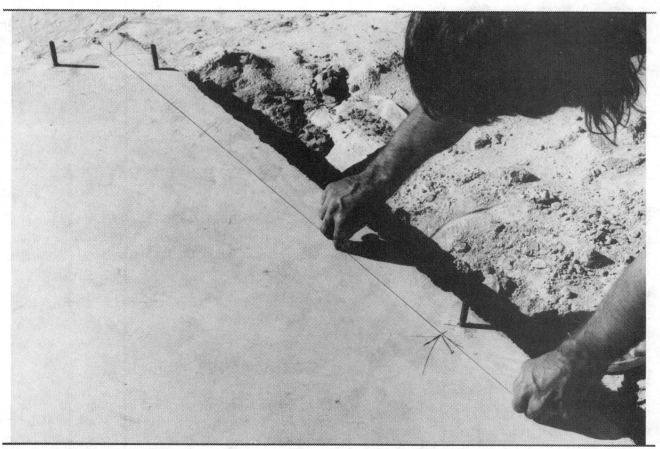

Figure 4-8   The wall line must pass through the intersection of the 5-foot and the 4-foot marks for the two walls to form a right angle. If this wall is very long, I recommend building a larger triangle for a more accuate snap.

Figure 4-9   Snap the long hall walls parallel to the front exterior wall. Then measure the perpendicular running walls off of the wall you squared with the front wall. That ensures a square building. Drive 1½-inch Teco nails into the slab or use a lead weight to hold one end of the string if you're working alone.

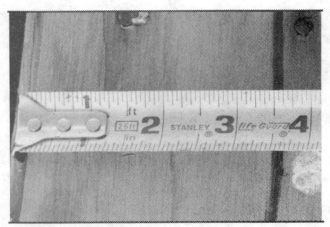

Figure 4-10   When snapping out interior walls, you'll usually measure from the outside of the building. By "burning" 3½ inches on the snapped line, you're measuring from the exterior of the building.

A lot of plans show distances measured from the outside of exterior walls up to the interior wall. In snapping lingo, this is called *"over, to it."* In other words, you measure "over" the exterior wall "to" (as opposed to over) the interior wall. When you lay out your tape, line up the wall line with the 3½-inch mark. That ensures that you're measuring from the actual exterior of the building. This is called "burning" 3½ inches (Figure 4-10). When you measure in to the interior wall, make a mark according to the number the plan gives you and then another mark 3½ inches past it to mark the other side of the wall.

■ Snapping Split Levels

Sometimes you'll run across split level slabs that require transferring a wall line from an upper floor down a step to a lower floor. Start by snapping out the wall line on the upper floor. Then transfer the line down the step with a level and your pencil to get one side of the lower wall established on the lower floor.

Now, hook a line on a Teco nail set in the slab on the far side of the line you snapped on the upper level. Have one person hold the chalk line against the level at approximately the height of the step and at the far end of the lower level wall, which is the side you have yet to establish (Figure 4-11). Now, with one person at the step guiding the person with the level, you can establish the other end of the lower wall. When the bubble is level and the chalk line matches the upper wall, the person with the level marks where the level rests on the slab.

When plating this wall, the bottom plate will obviously have to be cut, but the top plate should be one continuous piece.

■ Transferring the Details

When you've snapped the chalk lines for all the walls, begin transferring the location of details (doors, windows, posts, fire blocks, etc.) on the slab. This is done with a *keel*, which is something like an oversized oil crayon. It's better than a pencil because it stands up a lot longer to the rigors of weather and framers scuffing their feet over your numbers.

To detail out a window, write the components in the order they appear from the top of the window on down: the upper cripples, the header, the subsill (abbreviated SS), and lower cripples (Figure 4-12). Write this adjacent to the location where the window will be installed. Here are the components for a typical window: the top cripples run vertically from the top plate down to the header, the structural member which runs horizontally and carries the weight over the window opening: the subsill (not to be mistaken with a sill plate) is the horizontal member that the window's finish sill rests upon: the lower cripples run vertically from the subsill to the bottom plate and support the weight of the window.

A typical window detail written on the floor would look like this:

$$7\tfrac{1}{4}$$
$$\overline{}$$
$$63\tfrac{1}{2} \;\; 4 \times 8$$
$$\overline{}$$
$$60\tfrac{1}{2} \;\; SS$$
$$\overline{}$$
$$27$$
$$\overline{}$$

That means the top cripples are 7¼ inches long, and the 4 x 8 header is 63½ inches long. The subsill at 60½ inches is exactly 3 inches shorter than the header to allow for a 1½-inch trimmer to support the header on either edge. Finally we have 27-inch lower cripples below the window opening. Since the subsill is 60½ inches, this is the rough opening width for the finish window.

Mark the header size and upper cripples for doors in the same order. The trimmers are never detailed unless they're a special

size or will be left out altogether for some reason. Here's an example of a door detail:

$$7\tfrac{1}{4}$$
$$\overline{}$$
$$35 \;\; 4 \times 6$$
$$\overline{}$$
$$84 \;\; T$$
$$\overline{}$$

Again the top cripple would be 7¼ inches tall, and the 35-inch header would leave room for a 2′6″ wide door (30 + 3 inches worth of trimmers + 2 inches worth of door jamb and slack to level it). The trimmers are special since they're marked. Special means they're different from the typical door.

Figure 4-11 Transferring a wall line from one level down to another takes a little teamwork. First snap out the wall on top. To find the other end, drive a Teco nail in the slab, and pull a string to the other end of the lower wall. When the bubble reads level at the same time that the middle man says you're on the line, make a mark on the slab. The carpenter in the background is holding the string against the level. Once he gets the level reading true, the carpenter in the foreground will tell him how far off he is and in what direction he needs to go. He'll move over a little, level up and wait for his middle man to talk to him. Once the level reads true and the string matches the upper snapped line, he'll mark the slab where the level rests. Finally, snap a line between the two marks.

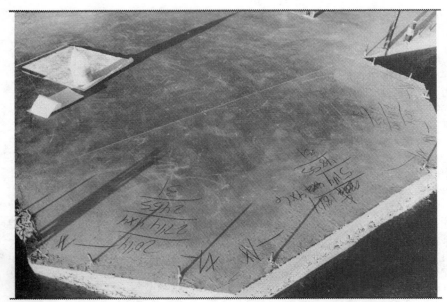

Figure 4-12 Window and door details should be written directly on the slab. For example, the center window has 18¼-inch top cripples, a 51¼-inch 4 x 6 header, a 48-inch subsill, and 31-inch lower cripples. The king studs were also located to simplify detailing the plate later. The keeled X's designate king studs. The keeled slashes designate a header. The keeled V is called a "chicken foot." Rather than just make a line, a carpenter will use a chicken foot to show an exact location. The point of the chicken foot designates the exact spot.

If you're working on a cement slab, the bottom plate will more than likely be pressure-treated lumber or green board (because it's slightly green from the chemicals used). Concrete slabs that sit directly on soil draw moisture up through to the surface. That can rot out the bottom plate if it's not pressure-treated to withstand the moisture, and the organisms that the moisture attracts. Slabs that are elevated (with parking beneath, for example) don't have this moisture problem. Some builders will permit untreated fir or pine on the bottom plate. This isn't something you'll find detailed in the plans. It's usually the builder's preference.

Remember that I've been talking about transferring the measurements from the plans to the slab. Some framers leave out this step. They prefer to wait until they have the plates cut and in place, and then detail directly on them. I like to put it all on the slab. Then when the plating is complete, I can come back, leave the plans in the truck, and simply transfer all my notes from the slab to the plates.

Cut one top plate and one bottom plate for each wall. To keep the pieces from shifting around during the detailing, many framers temporarily nail the top and bottom plates together with a few 8d nails. Later, when the walls are being built, they pull the plates apart and nail in each stud with two 16d nails. If you take the time to lay out and detail them all at once, you can be sure that every plate will be tight when you build the wall. That means the walls as a group will stand straight and level when they're finished.

Plating

Once the walls are snapped out, it's time to do the *plating*. The plates are the 2-by material that run horizontally on the top and bottom of each wall. Walls include three plates: one bottom plate and two top plates. The plating includes cutting, drilling, and placing the bottom and only one top plate. *Bottom plates* are sometimes referred to as *sill plates* or *sole plates*.

Here's a question that might occur to some readers: If there are two top plates on a wall, why not cut them both now? There are three reasons. First, the two top plates on a wall are actually two different lengths. On each corner, the lower of the two top plates (on a 2 x 4 wall) will be 3½ inches longer or shorter than the upper one. They're staggered at the corners to tie the tops of the walls together. Second, the lower top plate length is critical. It must be accurate to within ⅛ inch so it doesn't cause any adjoining walls to be pushed or pulled out of level. The length of the upper top plate is much

Figure 4-13   When spreading your plate stock, lay out as many pieces as possible before making any cuts. Place one end where it belongs and let the other end run wild. Now is a good time to mark out channels while you have the board in position.

less critical. Finally, you only need one top and one bottom plate to do the detailing. A second top plate at this stage would just be in the way.

Making a Square Cut by Eye

When plating, most experienced framers can make a square cut on the plating lumber without marking a square line across the face of the board. That saves time. You can keep your tape, pencil, and square in your bags. Every time you reach for a square, you lose time. It takes practice to make a square cut without using a line drawn on the board, but I recommend that you develop this skill.

The trick is to watch your saw table to be sure the saw's entering the board at a 90-degree angle. In time you'll be able to tell if it's square or not. I like to mark the cutoff length in the middle of the board. Then I drop the blade onto the mark, always pushing forward so the saw won't jump back. I keep

an eye on the table to be sure I'm entering the wood square. This works better than just entering the wood with your table flat. You have more table across the wood to sight with. Also, your blade isn't buried in the wood yet, so it's still adjustable if you see that it isn't quite square.

You'll keep up with the best of them if you learn to cut ends square without drawing a line on the board. Just remember to keep that forward pressure. And never move backward with a saw. It'll land in your lap before you know it. The circular saw causes more injuries on the building site than any other piece of equipment. Treat it with respect.

Laying Out the Channels

Begin plating by laying out the lumber, placing one end exactly where a wall stops (or runs into another wall) and letting the other end extend past the marked end point for that wall (Figure 4-13).

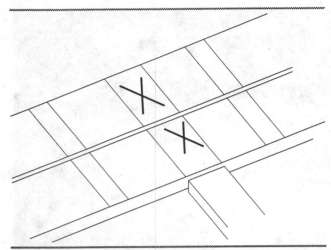

Figure 4-14 When walls intersect perpendicularly, a *channel* is required in the wall that receives the second wall. Detail a channel with lines that match the snapped lines on the floor. A keeled X designates the channel.

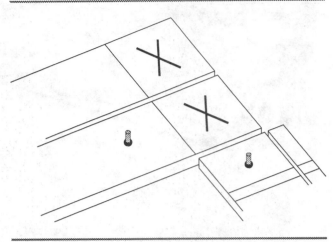

Figure 4-15 When two walls meet at a corner, the wall containing the channel is said to be *running through*. If you run one end of a wall through, always do the opposite end the same.

Usually I let the longest walls run through on the corners, as they'll be built first. When two walls meet at a corner, one will make up the corner itself, and the other, perpendicular wall will butt into it. The corner that's established is known as a *channel* (Figure 4-14). The wall with the channel on its end is said to be *running through* (Figure 4-15).

Always run parallel exterior walls the same way. If you run one through with a channel on either end, then run the opposite parallel wall through also. Also, never run one side of a wall through, and then hold the other end back. This is called log cabin plating, and it causes headaches when you go to raise the walls. Always be consistent with your channels.

To cut off the excess plate length, sight down to the intersecting wall line that's snapped on the slab. Make a shallow cut in the board where you judge the wall should end. Leave it a little long, if anything. Lay the plate back down and check by eye to see how it lines up with the line on the slab. Make any adjustments needed, and then cut the plate off at the correct length (Figure 4-16). This is a lot faster than taping, marking, and cutting the plate's length in separate operations. You're using the saw blade to mark the board, rather than your tape, pencil and square.

Figure 4-16 Now go through and make as many cuts as possible. I like to take this one step further by laying out material for the top plate next to each bottom plate. Then I cut both at the same time.

Figure 4-17   After the bottom plate is cut, lay it on some top plate stock. Make sure the ends are matched. Then trace the plate with the saw blade. This time-saver lets you keep your tape in your bag. That's where it belongs when you're plating.

By laying the board exactly in place on one end, you can eyeball the other end for length when you do the actual cutting. Lay it back down and see how close you got. If it's way off, try another board. Remember, this is called rough framing. Leave all the fuss to the finish carpenters! If you practice cutting by eye, in time you'll be making twice the money as the boys that insist on fumbling with their tapes and squares.

Continue throughout the whole house until you've cut the bottom and top plates for all the walls. When cutting the second plate, try placing the first one on top of a long piece of uncut material and tracing its two ends with your saw (Figure 4-17). Since the blade isn't large enough to go through two pieces of 2-by material, it will only

mark the lower piece of wood. Then you can pick it up and finish the cuts. Any time you can leave your tape in your bags, you're going to save time.

■ Drill the Bolts in the Bottom Plate

After you've cut all the top and bottom plates, go back to mark and drill the holes in the bottom plate for the foundation bolts. Lay the plate down so that each end is in place. You can't set it exactly where it belongs because the bolts are in the way, so place it right along the snapped line on the interior side of the wall. Now, using either a square and a pencil, or a bolt marker, spot the location for each bolt hole in the bottom plate (Figure 4-18).

Figure 4-18 A bolt marker is another time-saver. The tool has a hole set 3½ inches from a pointed peg to use for 2 x 4 walls. The crescent shape on the end of the tool is 5½ inches from the peg, to use for 2 x 6 plate stock. First make sure your bottom plate is resting along the snapped line. Then simply drop the tool over the bolt and hit the peg with your hammer. Make sure the tool is resting square in relation to the plate.

Most larger building material dealers sell bolt markers. Several types are made, but most work the same way. The half crescent on the outside edge is for 2 x 6 plates and the inside hole is for 2 x 4 plates. Make sure that you're using the right hole on the marker!

To use a bolt marker, begin by setting your bottom plate along the interior edge of the snapped wall line. Make sure it's also in place end to end, within the two end perpendicular wall lines. If you have a 2 x 4 plate, slip the bolt marker onto the first bolt, letting the bolt go into the drilled hole. This hole is exactly 3½ inches from the peg that's welded to the marker. Your plate must be lying right along the snapped line or the peg will mark the plate in the wrong place. If you're resting the plate ½ inch from the line, you will be marking the hole exactly ½ inch off.

When the plate is in place and the marker is on the bolt, swing the marker left or right until it sights square with the plate. Give the peg a good smack with your hammer so the pointed peg pierces the plate slightly. That's your mark for the exact location to drill. As you move down the line from one bolt to the next, keep adjusting the plate so that it

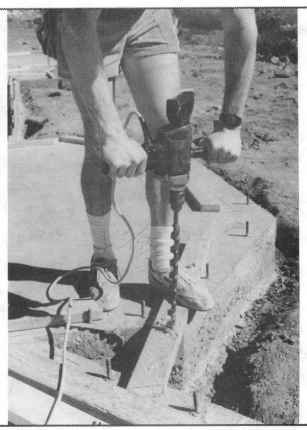

Figure 4-19 This Makita drill is the ideal setup for drilling plates. The drill has a trigger-activated reverse as well as the opposing handle. Both features are important for plating. The long bit, although expensive, is worth the cost. Do all your drilling at once. Check each plate to see that it fits the snapped layout exactly when laid over the anchor bolts.

lies along the snapped line. If your plate is even slightly crowned, you'll have to adjust it before you mark each bolt. It's not important whether the plate is along the line on any bolt you previously marked, or have yet to mark. Concentrate on the bolt you're working on, and how the plate looks in relation to it. Once the plate is drilled, it will straighten out as it's forced onto the bolts, if the bolts were marked properly.

I like to detail the 3½-inch channel mark on my bottom plate when I'm plating because the board's perfectly in place when I'm getting it ready to cut. It's easy to mark it then. Pull out your square and roll the snapped line right up and over the plating

Figure 4-20   If the bolts fit the holes, and the plate lies along the snapped line, temporarily nail up the top plate to the bottom plate and they're ready to be detailed. On interior walls you can stack the top plate on top of the bottom plate. On exterior walls such as the one pictured, the bolts make it impossible to temporarily nail the top plate on the bottom plate. Instead, hang it over the slab and nail it along the edge of the bottom plate. This still gives you two free surfaces (one for the bottom plate and another for the top plate) to transfer the detail onto.

material. If you don't do it then, you'll have to set it up all over again when you mark the bolts. It's only a matter of pulling out your tape and finding the 3½-inch mark. But multiply this by 50 or so channels and doing it twice — you get the point. Save time where you can.

After the bolts are marked, drill all the holes (Figure 4-19). Then slide the plate over the bolts to make sure it's a good fit. Making corrections later is a waste of time. Check it now when you're right on top of it with the drill in hand. If it fits, nail the top plate to it with a few 8d nails. Remember we're temporarily nailing the plates together so our detail markings match on both the top and bottom plate.

Don't put too many nails in the plates or the wall builders will be cursing you up and down. Two or three 8d nails is usually plenty.

Nail the interior plates together with the top plate on top of the bottom. Make sure the ends match perfectly before you send any nails. Exterior walls with bolts coming up through the bottom plate need to be nailed a little differently. Once you have the bottom plate down, the bolts hold it securely in place. Since the bolts are sticking up through the bottom plate, you can't nail the top plate directly on top of the bottom plate to detail it. Most framers hang the top plate over the edge of the slab, and nail

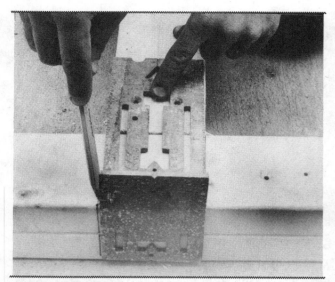

Figure 4-21   The last step in plating is to mark all the wall intersection points and channels. These need to be identified for two reasons. When the studs are being nailed to the plates, the framer will know to provide three studs here to make up the channel. Then, when he's cutting and applying the double top plate, he'll know to leave a space here for the intersecting wall's 3½-inch "ear" to lap onto.

it to the outside edge of the bottom plate (Figure 4-20). That way both plates are free to accept any detail. Again, the fewer the nails the better!

The last step to plating is to mark out all the channels with a square or a special channel-marking tool (Figure 4-21). That's the way framers communicate with each other that two walls intersect at this point. When two walls intersect, a channel is built into the wall for two reasons. First, the channel supplies a flat 2 x 4 so the intersecting wall can be securely nailed to the wall it's butting against. Second, a channel provides backing for the drywall on either side of the intersecting wall (Figure 4-22). Without channels there would be no way to tie the walls together where they meet. You'd also have to spend time adding backing for the drywall later.

To mark a channel, first make sure that the plates are in place in relation to the snapped lines. If they're in the right place, the points of intersection, or channels, will be easy to see. For a 2 x 4 wall, the channel detail will be 3½ inches wide and will follow the intersecting wall lines right onto the plates of the wall it's running into. This detail is simple and quick to make with a channel-marking tool.

A channel-marking tool is exactly 3½ inches wide and has two edges that are 1½ and 3 inches tall. If your plates are nailed together flat (on an exterior wall with bolts sticking up) use the edge of the tool that's 1½ inches tall. If the plates are nailed one on top of another, use the 3-inch edge. If you place the tool on top of the plates and in line with the intersecting wall, you need only trace the outside edges of the tool and you'll be detailing the channel. It's common practice to emphasize the channel detail making a large X with your keel inside the 3½ inch pencil marks.

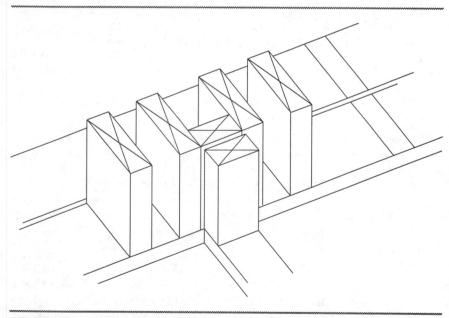

Figure 4-22   When a wall butts another wall, it needs only a single stud on the end to nail it into the channel built into the adjoining wall. The channel consists of a stud nailed flat that lines up with the end stud on the butting wall, and a stud set edgewise on either side of it to create drywall backing in the two separate rooms.

Figure 4-23   When detailing doors and windows, mark only the king studs. These are separated by the width of the header. This window is centered in the room (notice the "C" with the slash through it on the bottom plate). Since the header is 63¼ inches long, the king studs will be 31⅝ inches either side of the center. Detail men change their minds a lot, so you'll often see scribbles like the one by the king stud. Just ignore them. The dark streak leaving the king means that a header is coming off of this stud.

Keep in mind that when all the plates are laid out, it's easy to see the relationship of all the walls. Once the plates are detailed and the wall framers arrive, they'll begin by removing all the interior plates to make room to get the exterior walls up first. They might even build the interior walls out in the dirt surrounding the slab. If you don't detail a channel perfectly, how are they to know? If you missed it by ½ inch, the intersecting wall will be ½ inch out of square. Always make sure your plates are exactly in place before you mark the channels!

It's good practice to roll the marks from the top, down both sides or edges of the plates when possible. Having the marks along the edges of the plate is helpful in lining up the walls when they're being raised and nailed together.

When plating's done and the extra scrap wood is cleared away, it's time to start detailing.

Detailing Studs and Posts

If you were thorough in transferring detail marks to the slab before plating, then detailing the wall plates should be a breeze. Start with your windows and doors.

Mark on the plates the exact location of the king studs (the studs the header nails into). See Figure 4-23. Typically an interior door header will be 5 inches wider than the door. If you have a 2'6" door, your header would be 35 inches long. This allows

3 inches for trimmers, 1½ inches for door jamb (¾ inch either side) and ½ inch to level the jambs when the doors are going in. So if you're detailing a 2'6" door, lay out two lines on your plates 35 inches apart and emphasize that they're king studs by adding an X on the outer sides of each line. Also strike a line off either king stud mark with your keel to show that a header will be installed here.

I also like to note the header size on the top plate. Later, when the walls are being built and all the plates are shuffled around, it helps to have that one clue written plainly on the plate if you get confused. Go through the building and mark all the doors. The plans will give you exact numbers, or you might have to approximate each door location.

Figure window header sizes by adding to the rough openings provided by the builder or window company. For example if the manufacturer gave the rough opening for a 24 x 48 inch window as 24½ x 48½, the header would be 27½ inches long. (Window callouts are width x height.) All you do is add 3 inches (two trimmers) to the rough opening to find the header size. As a rule of thumb, any header over 7 feet long will need two trimmers on either side. So you would add 6 inches to the R.O. (rough opening).

I've seen plans that detailed (on the "structural" page) 4 x 4 trimmers to carry an extremely large overhead load. In that case, you'd add 7 inches to the R.O. to get the header size. Time spent reviewing the plans before you start detailing is time well spent. Window headers are detailed exactly like door headers. Use a straight pencil line for the king stud, an X made with a keel to establish that it is a king stud, a slash mark off the king stud with keel to show a header is going there, and finally the header size itself. If you have double trimmers or a 4 x 4 trimmer, make a note next to the king stud mark. Double trimmers are detailed with double T marks (TT). If you haven't already done so, write down the window components directly onto the slab with your keel.

Next, locate and mark out all the structural posts. Find their exact size and location and detail them directly onto the plates. If a 4 x 6 is shown holding up a beam that will eventually sit on top of

the walls, mark it down on the plates so it's included when the walls are built. Establish its location according to the prints, and mark it out on the plates. A 4 x 6 in a 2 x 4 wall would sit lengthwise — so you detail it with two pencil lines 5½ inches apart. Emphasize it with a keeled X within the penciled detail. Also write the size of the post in this space.

But don't all those X's get confusing? How can a rookie carpenter tell a post X from a king stud X from a channel X? The trick is in the pencil marks. A keeled X is there to shout "Pay attention, something's about to happen." When a wall framer is sending nails through the top plate into the studs with one swing per nail, his mind is usually far off. It takes a blatant, fat keeled X to snap him back to reality. The X will make him stop and ponder for a brief instant. Is it a king stud? If there's a header slashed off the mark, then yes. A post would be marked out exactly with pencil marks and the size of the post. An X only alerts the framer that there's something besides studs 16 inches on center here. The pencil marks explain exactly what it is.

If there are drop ceilings, it's much easier to put in the fire blocks (or draft stops) when the walls are being built. Fire blocks are required at all intersections where a drop ceiling connects with any wall. For example, a hallway may be dropped to 7 feet from the floor to allow the heat to pass down the hall. You would detail (in keel) on the plate "Blocks 84 to center." Then the wall framer would build in blocks between all studs with the block centers at 84 inches off the floor, hooking his tape on the bottom plate and going up 84 inches. Later when the drop was constructed, the blocks would be waiting, to the joy of the drop man. The pay for framing walls is the same whether the fire blocks are detailed or not. Drop ceilings will always cost more if fire blocks need to be installed.

Also, balloon and rake walls that have sheathing on the exterior require a block set every 8 feet of wall height to catch the plywood edges. A detail keeled on the plates reading "Blocks 96, 192 on center FF" (from floor) would take care of this situation. Buildings with wood siding often need to have special backing installed in the walls to catch

Figure 4-24 It's not necessary to keep the stud layout exactly 16 inches on center if the wall isn't a shear wall. Fit studs in between the king studs and channels. I like to leave a bay centered over each doorway in case the HVAC contractor wants to put a vent there. On tracts, you aren't required to lay out any cripples at all, unless the plans call out a heat bay centered over a doorway.

the edges of the siding or to give appropriate nailing. Figure out your installation procedure now and save some dollars later.

I recommend detailing posts, fire blocks, and any specialty items on both the top and bottom plate and on the floor adjacent to each appropriate wall. It might sound like overkill to note all of this in so many places, but there's a good reason. Once, I pieced out the walls on a project and not one fire block was installed, even though I had detailed quite a few of them. When I questioned the guy about my missing blocks, he said "What blocks? I don't see any block detail." Sure enough it was gone. He'd flipped over all the top plates that I had detailed so that the detail marks were hidden. If I had made detail marks on the floor, or even on the edges of the plates, he wouldn't have gotten away with it. Some folks would scoff at the

thought of detailing fire blocks. Yet I can guarantee you that it's a hundred times easier to install them while you're building the walls instead of from a ladder later on.

The last step in detailing is laying out the studs (Figure 4-24). The only place that it's important to keep a consistent layout from one end of the wall to the other is on a shear wall, or any wall that will receive plywood mounted on its exterior. The drywallers I've talked to say that since they run their sheets horizontally, it doesn't bother them to have an inconsistent layout throughout the interior of the building, except on long walls. Just make sure that you don't leave any bays greater than 16 inches.

Keep the layout consistent on a shear wall. (That's a wall designed to resist lateral, or shear, forces. It'll be covered with structural plywood and

Figure 4-25   Studs in a shear wall must be laid out exactly 16 inches on center. This shear wall is designated by the delta 6, which is the nailing schedule. Hold the layout bar half a stud (approximately ¾ inch) over the beginning corner to assure 48 inches to the center of the fourth stud. In this case the fourth stud is an upper and lower window cripple.

nailed according to the shear schedule in the plans.) I like to mark all the shear walls with paint right before I lay out the studs. I'll spray right on the slab the nailing schedule written inside a delta, or triangle. (But never spray a garage slab as it will more than likely be exposed!) The delta is meant to tell the installer on which side of the wall to apply the plywood. Exterior walls usually have the plywood on the outside. Interiors will vary. Therefore I like to use the delta to keep things clear.

Notice in Figure 4-25 the delta is pointing out, and it has a 6 inside it. In this situation the plywood would go on the outside and be nailed 6 inches on center along the edges and 12 inches throughout the field of the plywood. If the delta was flipped and pointing at the viewer, the plywood would be mounted on the interior. If you go through the

building now and label all shear walls and their nailing schedules, you'll never have to hunt them down in the plans again. If you don't, you'll be looking them up at least three more times throughout the life of the project.

A very important part of framing layout is the act of holding the first stud (or joist or rafter) back ¾ inch so a standard sheet of plywood will land in the middle of the fourth stud. If you hooked your tape on the outside edge of a wall and marked out a straight 16 inches O.C. (on center) layout, the fourth stud would land between 48 and 49½ inches. A standard sheet of plywood is 48 inches wide. So it would land exactly along the first edge of the stud. In order to prevent this, it's common practice to hold the first stud back ¾ inch so it lands between 15¼ and 16¾ inches. Then you can pull a straight 16 inches O.C. layout off of it. Then the fourth stud would land between 47¼ and 48¾ inches. The plywood will land exactly in the middle of the fourth stud. This seems like such a simple concept — yet I'm constantly amazed at the number of licensed framing contractors that have never grasped it.

A great time-saving tool for laying out studs is a layout bar, shown in Figure 4-25. This crafty device is made up of four short bars welded to another 49½-inch bar that acts as a handle. The shorter bars are exactly 1½ inches wide and spaced at 16 inches O.C. If you rest the bar on the plates and strike a pencil line along the edge of each of the shorter bars, you're marking out 1½-inch spaces that are laid out 16 inches O.C. That's the exact detail for studs.

After the exterior and shear walls are filled with stud layouts, move to the interior. At this point you'll have channels, king studs and special structural posts laid out on your plates. Fill in the open spaces between the king studs and the channels and posts with stud layouts. Start at the edge of a wall and lay out studs 16 inches O.C. until you come to a channel, king stud, or post. For doorways and windows, you can continue laying out right on through, between the king stud marks, as you'll need layouts for your upper and lower cripples. If you approach a channel and a stud lays out directly on the channel, don't lay it out! You would only be creating confusion by adding two more pencil lines. A channel is made up of three studs that have a precise location already laid out. It has priority because it is laid out precisely at an intersection. That's why I lay out studs last.

Fill in the open spaces between your details with studs, leaving no bays larger than 16 inches. Although it goes against the grain with the old school of framers, the straight 16 inch O.C. layout from one end of the wall to the other just isn't important. Of course if I have a 20-foot wall, I'm going to keep 16-inch bays along its length to accommodate the drywallers. But for short 6-, 8-, or 10-foot sections of bathroom and bedroom walls, I'll just fill in the spaces between any previous layout. Drywall comes in 12-foot lengths and is commonly run horizontally. One sheet will cover the whole wall.

Any interior wall that gets shear plywood will have a constant layout throughout its length, no matter what. Plywood is installed vertically, and needs every fourth stud to be perfectly on layout.

Detailing a Curved Wall

Radius (curved) walls are a common design element in new homes today. Even though lumber is intended to be straight, building a curved wall is easier than it sounds. First check the plans for the size of the radius. The radius in our example is at a corner where two walls meet. We'll scale the dimensions off the plans, which are drawn at a scale of ¼ inch equals 1 foot. Measuring the distance of the radius from one wall to the next on the plans, I came up with ½ inch. At ¼ inch per foot, that's a 2-foot or 24-inch radius.

Usually a radius wall is added to a house to round out a sharply angled corner. This adds a graceful flow to bends in hallways or corners of large rooms. If a house has only a few radius corners, they're probably all the same size. If a house has a lot of radius walls, they may vary in size, depending on the location and the effect the designer is trying to achieve. Scale off each radius from the plans to see if they're uniform or not.

Once you've established the size of each radius, you can cut your plates. Start by driving a Teco nail into the corner of a piece of ¾-inch plywood. Hook your tape onto the nail. Then, holding your pencil on the 24-inch mark, pull your tape around in a semicircle (Figure 4-26). Next, add 3½ inches and repeat the step to mark out the outside of the plate. To cut out the plate, set your blade for about half the depth of the plywood and slowly follow the pencil marks. Sometimes you'll have to pull the blade up and reposition it, but don't make a chopping motion with the saw. That creates ridges. Just cut very

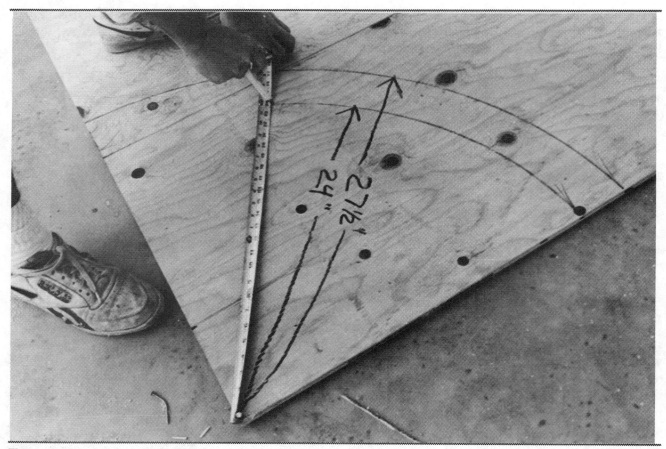

Figure 4-26   Scribing a radius plate is very simple. Put a Teco nail in the corner of a sheet of ¾-inch plywood. Drag your pencil around while holding it at the right number on your tape. Add 3½ inches to mark the radius for the outside edge of the wall. But be careful not to pull too hard on the nail. Bending the nail will put a bow in your radius.

slowly. When you've made one pass, set the blade the full depth of the plywood and pass it through again to complete the cut.

When cutting radius plates, the saw blade will heat up from the tension of the unnatural bending you're forcing on it. When it reaches a certain temperature, it will start to warp and wobble. What do you do? Keep the nose of the saw planted in place and lift the rear end up. This will in effect free the blade from the wood and allow it to spin freely in the cool breeze. Run it for a few moments, or until you see the blade straighten back out. Then drop it right back in where you left off. The tighter the radius, the harder it is on your blade. A few inches down the radius, if it heats up again, leave the nose planted, lift the rear, cool the blade, and proceed.

But remember this. When you lift the blade out of the wood, pay attention. Never unlock your wrist or elbow! Respect that circular saw and leave your jig saw in the truck. I've yet to met a radius plate I couldn't cut with a circular saw in half the time and with the same precision as a jig saw. It just takes practice, patience, and respect.

When the first plate is cut, use it as a pattern and cut five more. Using 1¼-inch Teco nails, nail the plates together in groups of two. You'll have a 1½-inch bottom plate and two 1½-inch top plates, which allows you to use the same size studs as your other walls.

Now try fitting them in place. If they're too tight, trim the ends (Figure 4-27). Using a small block, lay out the studs (Figure 4-28). I'm often

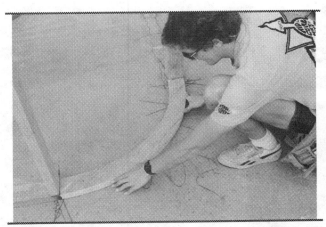

Figure 4-27 After cutting out the plates, check that they fit lengthwise in the allotted space. Do any necessary trimming.

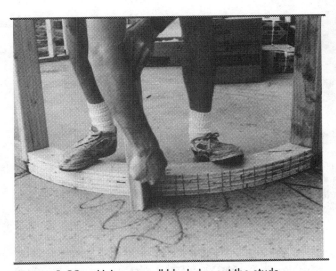

Figure 4-28 Using a small block, lay out the studs.

Figure 4-29 Cutting headers and cripples is easy with a radial arm saw. Even a small unit like this one will make quick work of a basic wall package of subsills and cripples.

accused of overkill on the number of studs I put in radius walls. But the more studs you have, the easier it is to bend the drywall and the cleaner the finished wall looks. If you put in too few studs, your radius wall ends up looking like a series of flat surfaces instead of a true radius.

The last step in snapping, plating and detailing is to cut up and spread all the headers, cripples, subsills, and posts. On small jobs, it's just a matter of making a list of everything that's detailed and cutting it all up at once (Figure 4-29). List all your cripples (top and bottom), subsills, headers and posts. Set up a cut area and cut all the pieces. Finally, pile the cut pieces next to each door and window (Figure 4-30). Some framers like to nail together all the channels as well. Others prefer to cut the cripples and subsills as they're building the walls. This works, but it takes a little more time. When you're building walls, you want to blow and go. Messing around with cutting cripples can take the wind right out of your sails.

The Pay

On a tract you'll find large yards set up just for cutting. A framer will get a set amount of money ($20 to $50) to cut up the package (the group of

Figure 4-30   If slabs are poured fast enough on a tract, you can make good money on snap, plate and detail work. Once you've done a few units, you start to memorize the measurements. Then you can really pick up speed. Notice the cripples and headers spread out in place. Some upper cripples have already been nailed to the headers and are standing up.

headers, cripples, subsills, and posts for each unit). These packages are labeled, bundled, strapped and dropped at each unit, once the snap, plate and detail is finished. Some tract jobs require the snap, plate, detail crew to spread out the header, cripple, and subsill packages, leaving the header and cripples adjacent to the doors and windows.

The flow of work will vary among custom jobs. If there's enough time, someone will cut up the entire package. But you may have to cut cripples as you frame. I prefer to cut my packages first because it gets the wall done faster.

When this was written (in 1993), snap, plate and detail work was paying from 9 to 20 cents per square foot of building. Since plating is the labor-intensive part of this job, you'll usually find an experienced carpenter snapping the walls out on the slab while a laborer cuts the plates. Then the snapper will drop back and detail what's been plated. Two people should average about 3,000 square feet of building a day once they're on a roll. And remember, you never count garages when figuring square feet. Some ageless unwritten rule makes us build them for free. Don't ask me why!

FIVE

Building the Walls

When you mention framing, most people think of wall building — banging studs together and lifting long walls. That's the impressive part, since it looks like the house goes up in a day. In reality, however, building the walls is a relatively minor and unskilled job. Most all of the brain work was taken care of by the detail man.

The Planning

Building the walls is like working a puzzle. The detail man has marked out exactly where he wants each window, door, channel, post, block and stud in the building. Once you've learned to read detail, it's a piece of cake to bang together the entire building, one wall at a time.

Except for learning to read the detail on the plates and a few tricks for speed, anyone can pick up the basics of wall framing in a day. Because it's labor intensive (as opposed to brain intensive), it doesn't pay as well as other framing jobs. But it's a great place to start if you're just beginning in the trade. The theories of basic layout are the foundation of all rough carpentry — and most everyone learns them through wall building. If you can build a wall, you can build a floor. If you can build a floor, you can build a roof, and so on.

Begin the job by studying the detail on the plates and clearing up any questionable areas with the framing foreman or detail person. Then decide on your plan of attack. The trick is to figure out which walls go up first. Begin with the longest exterior walls that have their channels

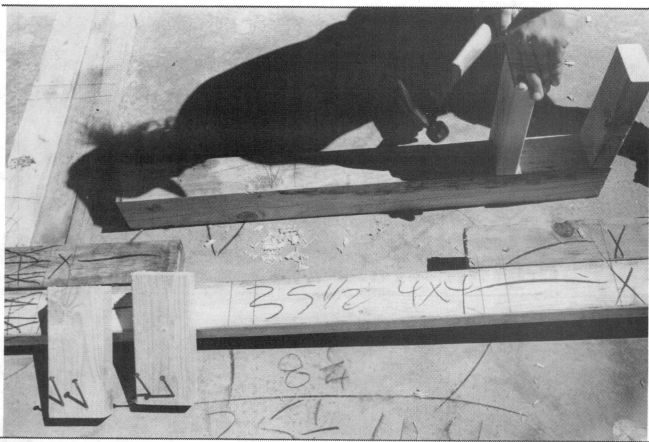

Figure 5-1   Before beginning any wall, *buck up* (nail together) all the cripples to the headers and subsills. Spread out all the items that are detailed where they belong. Lay the headers right on the plate and transfer the layout from the plate to the header. Then toenail the cripples to the header as shown.

running through on the corners. Usually the front and back walls are the longest, so you'll do them first. Then frame as many other walls as you can fit on the slab at one time. Follow up with the sides and any long interior walls that run parallel to them. Finally, frame the small interior walls (usually called the *guts*).

The trick here is to keep from framing yourself into a corner. When work space on the slab is limited, there may not be enough room to build your wall. Or worse yet, maybe you framed a wall where space was available. Then you discover there's no way to move it to where you need to raise it. My advice: do a little planning. Take the time to visualize the direction that individual groups of walls are laid out. Walls running in the same direction should be framed at the same time, if room permits. Always build as many walls as possible. This is a very simple — yet very important — point. Building one wall at a time is inefficient. An organized framer can build a typical 2000 square foot house in three to four lifts (groups of walls).

The Building

Once you've decided which walls to build in your first lift, precut all the headers, the top and bottom cripples and the window subsills for that lift.

Figure 5-2   You'd build the exterior wall to the far left first because it runs past the back wall by about a foot. You could frame it out in the dirt, which would give you enough room to build the front and back wall at the same time. If you don't feel comfortable remembering where each interior wall goes, label each one before moving it out of the way.

Spread out these cut pieces on the slab where they belong. On some projects everything will already be cut and spread for you.

Beginning the Walls

Start building by nailing up your cripples to the headers and subsills. Figure 5-1 shows the process. Lay your headers and subsills directly on the detailed plates and transfer the layout from the plate to the headers. That's faster and easier than taking out your tape and dealing with a bunch of unnecessary numbers.

Confirm whether the foreman wants the studs crowned or not. Some companies want the crown in all studs to be in the same direction so the walls are straight (or at least bent in the same direction). Some don't bother. Then you precut all studs for that lift, including any balloon studs (studs that are two stories tall).

To make efficient use of your working space, move the interior plates off the slab so that you have room to build the exterior walls. If necessary, put an identifying number on each plate. Put the same number on the slab where the plate will go. That helps keep confusion to a minimum.

Figure 5-2 shows the floor plan plated and the cripples nailed to the headers and subsills for the first floor. On this building, the wall in the left foreground and the opposite wall in the right background would be built first because they're the

Figure 5-3   Spread out the studs, headers, and subsills for as many walls as possible. In this figure an exterior and a long interior hall wall are being framed together.

Figure 5-4   Here we have a window all spread and ready to nail up. The studs, the header and the subsill setup are laid out according to the detail on the plates.

longest running exterior walls. Move the interior wall plates out of the way and start spreading a stud for each layout mark detailed on the plates (Figure 5-3). Next, load up on nails and start nailing along the bottom plate of your first wall, nailing each stud in place with two 16d nails. Try to continue in one direction as long as possible.

If you have a few walls spread side by side, nail all the bottom plates before moving to the top plates. Moving from top to bottom on each stud wastes time. When you come to a window or door (Figure 5-4), nail up the cripples to the top or bottom plate and the header to the king studs and continue on. (The king stud is the stud that the header is nailed to.) Later you'll cut a trimmer that will nail up along the king stud and support the header (Figure 5-5).

Figure 5-5 Cut door and window trimmers just like you cut blocks. Lay two studs in place. Then sight your cuts along the header (in this case a 2 x 4). Always cut a little long and then adjust if needed. In this photo, the trimmers have been cut and laid in place, with the right one ready to be nailed. The cut-offs could be used for the blocks on each side of the door.

■ Cutting Blocks, Trimmers and Plates by Eye

The first step to cutting is to get material spread out for all the items that need cutting. Depending on the area where you're working, you might need corner blocks, door blocks, trimmers, medicine cabinet blocks, fire blocks for drop ceilings, let-in braces, and finally the double top plate. Do all your cutting at the same time, and then go back and nail it up. With a crew of two, one person should do the cutting, and the other should follow up, nailing all that has been cut.

To get good (and fast) at wall building, you'll have to learn to cut your blocks, window and door trimmers, and top plates by eye. There is rarely an excuse to take out your tape when building walls. It's a waste of time. Figure 5-6 shows how to cut corner blocks by eye. Lay out some material along the top or bottom plate of the wall. Never cut blocks in the middle of the wall as the studs may bow slightly. Butt your material up against the inside

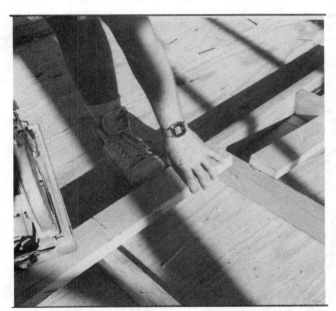

Figure 5-6a To cut blocks quickly, butt a piece of scrap up to one side of the bay you're cutting. Sight down on the opposite side and make a shallow cut in the wood with the saw where you think the block should be cut.

Figure 5-6b Move the saw aside and see if cutting at this mark will make a tight block. Then make any adjustments necessary when you do the cutting. In time you'll develop an "eye" for accurate cuts, making just one cut for a perfect block every time.

Figure 5-7a   Let-in braces should span five or six stud bays (on an 8-foot wall). That makes the brace run at about a 45-degree angle. On long walls, set a brace about every 20 feet. Lay the brace in place and set your saw blade depth to a little more than 1½ inches. Cut into the studs along the sides of the brace. Then cut the brace top and bottom flush with the plates.

edge of the first bay. Look down at the inside edge of the next stud and sight down to where you feel the block should be cut. Make a slight cut at this point.

The first photo shows that a slight cut has been made and the carpenter is now sighting at how close his guess was. If this cut was made, would the block fit, or would it be too long or short? Now is the time to adjust. Make sure to sight that the blade is entering the block square. The next figure shows that the cut has been made, and the block is now being checked in the bay for size. If it's slightly long, cut it down to fit. When you're first learning this method, it's best to make your cut a little long and then trim it down to size.

Corner blocks are installed in the outside three bays of exterior walls. They're cut to length, the width of each bay. This helps stabilize the corners of the building. Door blocks are installed in two bays on each side of every door. Corner and door blocks are set at approximately mid-height up the wall. Trimmers are installed against the king studs and underneath the header. Medicine cabinet blocks (typically detailed on the plates with a keeled MC), the width of the bay, are installed 4 feet and 6 feet above the floor. This leaves a 2-foot vertical bay.

A let-in brace is a brace that holds a wall plumb. It's usually a 1 x 6 installed at approximately a 45-degree angle. Why is it called a let-in brace? Because it is actually cut in (or let in) ¾ inch on the edge of each stud. See Figure 5-7. Fire blocks are

Figure 5-7b   Reset your saw blade to full depth. Holding the saw sideways by the front of the table and the back handle, drop the blade into the wood as you move the saw forward. You'll be cutting at least ¾ inch out of the stud, from one scribe mark to the next. If you don't cut low enough, the brace won't be let far enough into the wall, causing the drywall to bow. This looks dangerous, but it's actually a very controllable cut. Remember to keep that forward pressure constant. *Never* try to move backwards.

Figure 5-7c   Set the brace in place and nail it to the bottom plate and the first stud with three 8d nails on each. Only nail the bottom plate and the first stud. The rest of the brace will be nailed off completely during plumb and line. Tack in three 8d nails in each of the following studs. Bend a few nails over the top end to keep the brace from flapping around when you lift the wall.

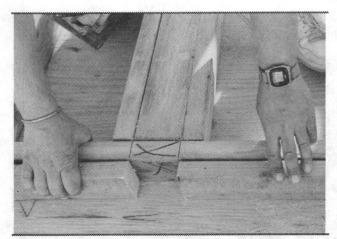

Figure 5-8   When cutting the double top plate, leave all the channels open to accept *ears* (3½ inch overlap) from perpendicular walls. If the end of the wall has just one stud and no corner channel, leave an ear to lap onto the wall it meets.

cut the width of each bay and their height is either detailed on the plates or determined at the time of installation. The double top plate is installed directly on top of the top plate.

■ Doing the Nailing

Nail on your double top plate, leaving the channel open (as in Figure 5-8) to receive the end of the top plate (ear) from the intersecting wall. Figure 5-9 shows a good trick for cutting the double top plate. By cutting the top plate on the opposite side of the channel, you allow the 3½-inch ear to extend over the end of the wall. When nailing, hold the top plate

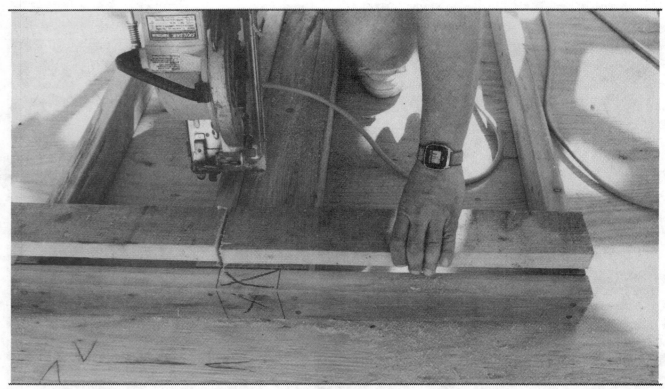

Figure 5-9   Here's a good trick for cutting the double top plate. If you have a 3½-inch channel in the middle of the wall and a 3½ inch ear going over the end, lay down your material so that it's flush with the side that gets the ear (the right side in this photo). Now cut on the opposite side of your interior wall channel. When you open up the channel 3½ inches, you'll have the 3½ inch ear you need. If you nailed the top plate as shown here, you would block the channel from accepting the intersecting wall's top plate ear. Only use the channel mark for a 3½-inch guide. Cut on the left side of the mark, but nail it on the right hand side (as shown in Figure 5-8) in order to have a 3½-inch opening and end ear.

Figure 5-10   After the wall is lifted, temporary "A" braces hold it in place until interior walls stabilize it. Notice the center of the fireblock line is at the top plate of the normal studded wall. When the rooms are joisted, these blocks keep fire from passing up between the bays and into the attic or second story. This is a shear wall that ties into the roof, or it would have required let-in braces every 20 feet along its length.

along the right side of the channel mark (Figure 5-8) and let the extra 3½ inches of top plate extend over the lower top plate to tie the two intersecting walls.

On walls that won't have shear panel installed, you'll have to cut in let-in braces. See Figure 5-7. Be sure to span at least five stud bays.

Raising the Walls

When you're ready to tip the wall up, sweep the slab where the bottom plate will rest so you don't trap anything under the wall. Raise the wall and put up temporary braces to hold everything in place while you do the next lift. Figure 5-10 shows a wall that's temporarily braced.

Once the exterior walls are up, frame up your interior walls. Remember to build as many as your floor space allows. Figure 5-11 shows four room partitions being built at once. Earlier we spoke of

Figure 5-11   Once you've built the exterior and any long interior walls, start with the small interior walls. Build as many as you can on floor space that's available.

visualizing groups of walls that would be built together. Here is a classic example. By nailing up all your studs and then cutting all at once, you can build four walls in the same time it would take to build one.

When building walls, always build and raise your tallest balloon and rake walls first. Otherwise you may not have room later. Figure 5-12 shows an entry way with first floor walls that balloon to match the top of the second floor walls. These walls will eventually be surrounded by normal 8-foot walls.

■ Using a Crane to Raise the Walls

Large balloon walls may be too heavy and awkward to raise by hand. A crane or rough terrain forklift (pettibone) may be required. See Figure

Figure 5-12   Build any balloon walls early, while there's still floor space available!

5-13. Most stairways and vaulted living rooms require first-story top plates to continue to the height of the second-story top plate. When the second-story walls are built, their top plates will match the top plates on these balloon walls. The roof will then stack evenly, all the way around.

Trying to raise balloon walls by hand can be very dangerous. They may look innocent enough on the ground, but half way up you may realize they're too much to lift by hand. Lowering a wall back to the slab without doing any damage to yourself is difficult. If at all possible, get a lift for the balloon walls.

If there's room to leave balloon walls down, I like to get all the lower 8-foot walls built, plumbed and lined, have all the ceiling and floor joists organized, and then call in a lift. The operator can raise all the balloon walls and all the joists at once. But on some jobs you have to raise the balloon walls before you can build the normal height studs. It's best to have a lift available all during the framing if that's financially feasible.

■ Topping Off

The last step in framing walls is called topping off. That's when you tie together all the ears on the double top plate. This involves nailing the laps on the top plates, as shown in Figure 5-14. Use toenails to tighten the gaps, and then put two 16d nails into the plate that's lapping. If you haven't got the balance to walk top plates, use a ladder! Next, check to be sure that all the channels are nailed up. Where two walls intersect at a right angle, drive six nails connecting the channels: two on the top, two in the middle and two on the bottom. Finally, walk

Figure 5-13   A rough-terrain forklift is great for lifting balloon walls. When the wall is about a quarter of the way up, tack some long 2 x 4s to the wall. Use these "handles" to guide the wall into place. Once the wall is up, use the same handles as temporary braces.

through the building and double-check all your nailing. You can bet the guy signing your check will do the same.

Building Rake Walls

A rake wall is angled at the top, usually matching the pitch of the roof or ceiling. Sometimes a wall has normal studs up to a point, and then it takes off as a rake wall. This change will be detailed on the plates as in Figure 5-15. In this figure, the rake wall starts on the right side of an interior wall channel. Notice the arrow points to the exact stud that will be the long point of the rake. The slab is detailed in

keel with the notation, Stud LP 165. This means the stud (not top of the wall) will be 165 inches at its long point. The opposite end of the rake wall will detail the short point of the wall. If you're lucky, the snap, plate and detail crew will have snapped out this angle on the floor already.

In most cases, you won't be provided the top plate on a rake wall. That's because it will end up much longer than the bottom plate. Usually, only the bottom plate is detailed. You'll cut the top plate after you've built the rake wall.

Begin by spreading out all the common studs up to the point that the rake begins. For the rake studs, lay out a long 2 x 4 on each stud layout.

Figure 5-14   Before you top off a wall, check to make sure that the lower top plates are fitted together tightly. If the plates aren't tight, the walls won't plumb up consistently. Try angling a toenail from the lower plate up into the ear, as pictured. This will help snug up the plates before you drive two nails down from the top. Each ear gets two 16d nails.

Figure 5-15   This plate detail shows a rake wall adjoining a normal height wall. The backing stud on the right side of the channel will be the tallest point of the rake wall. (Notice on the bottom plate the word "rake" and the arrow ending on the right side of the channel.) This stud will measure 165 inches at its long point, which is called out on the slab.

Make sure it's long enough to reach the required height. See Figure 5-16. Nail each stud to the bottom plate.

Because a rake wall continues up past the ceiling height of adjoining rooms, you'll need a fire block in each bay. Cut your fire blocks now and use them to help find the lengths of your rake studs (Figure 5-17).

Temporarily set a fire block in each bay, up toward the top of the wall. Check to see that the bottom plate is aligned with its snapped line on the slab. Now mark out the short and long points on the two end studs and snap a line from one to the other, as shown in Figure 5-18. If the end studs are set up at exactly the same width that's detailed on the bottom plate, the fire blocks will hold each of the remaining studs on layout.

Figure 5-16   First spread out and nail up all the normal height studs. Then spread studs that are at least 165 inches long for each layout on the rake section of the wall. Nail the studs to the bottom plate.

Figure 5-17   Cut a fire block for each bay of the rake section. Set them up towards the top of the studs, as pictured in Figure 5-18.

Figure 5-18 Measure up from the bottom plate and mark the long point stud and the short point stud. Make sure that if you hook the bottom plate, add 1½ inches to your stud length. Check the bottom of the wall to see that it's perfectly straight along the snapped line on the slab. Now snap a line on the studs, from the long point to the short point.

Figure 5-19 Using your square, mark out the long point on each stud. Depending on the direction you build your rake, the table on your circular saw may not swing in the right direction for you to be able to make the angled cut on the stud. But don't worry. A good trick is to cut the long point square first, and then make another cut following the angle of the snapped line on the edge of the stud, while sighting the square cut you made at the long point.

Transfer the snapped line down the face of each stud, starting at the long point. Cut the stud square at this point (Figure 5-19). Then turn your wrist slightly until the blade on your saw matches the snapped angle. Make another cut, keeping the saw angled with the snapped line and following the long point square cut you made previously. Or, depending on the direction of the rake, you might be able to use the swing on your circular saw table to cut the angle. Since the saw table only swings one direction, you may not be able to swing your table in the right direction. If you own a Makita Sidewinder saw, you're in luck! The table on this saw swings in both directions.

Once all the tops are cut, nail up the fire blocks at the height of the common top plate as in Figure 5-20. Check to make sure that the wall width is consistent from top to bottom. Sometimes the fire blocks are a little short and the wall ends up skinnier on the top. If this happens, hit a few studs near the fire blocks to widen the top. When you have a consistent width, cut two top plates and nail them to the top of the studs. Raise the wall and temporarily brace it off as shown in Figure 5-10.

Many rake walls have one long point and two opposite short points, as in Figure 5-21. Build these just the same way; nail up the bottom plates, cut the

Figure 5-20　Nail up the fireblocks at the height of the common top plate. Check that the bottom plate is straight along the snapped line on the slab and then cut two top plates and nail them on. In this photo, the first top plate has been cut and nailed on, and the second plate is ready for cutting.

Figure 5-21　Rakes with one long point and two short points are built exactly the same way.

Figure 5-22   When you nail together radius walls by hand, try setting all nails into the plates first. Then drive the nails snug into the studs.

fire blocks, snap and cut the tops of the studs, and finally, cut and apply the top plate. You would definitely have to build a rake this size before any other wall.

Building Radius Walls

Radius walls are framed like any other wall. If you're nailing them by hand, it's easier to set all the nails in your plate first (see Figure 5-22) and then bang them in. Raise the radius wall section and then drive two nails top, middle, and bottom, into the two end walls it leaves from, or about every 4 feet if the wall is more than 8 feet high. Figure 5-23 shows a short radius wall.

Can You Make Money Doing It?

While wall framing isn't the highest-paying carpentry job, a well-organized crew can make good money. It just takes planning and organization. There are two tricks: Cut down on the number

Figure 5-23   Radius walls make a nice design feature. They really help hallways flow smoothly from one room to the next. You'll find radius wall in more and more better-quality homes. If you use untreated plywood for the bottom plates, slip some black roofing paper between the plate and the slab. This helps prevent rotting in the bottom plate.

of steps you take and build as many walls as possible in each lift. To be most efficient, stick with each step, completing it before you continue on to the next. Wall building generally pays from 15 to 60 cents per square foot of floor. Jobs with cut-up walls and highly-customized designs always pay more. You might earn $400 to frame a typical 2,000 square foot tract house — and three framers could power it out in a day.

Plumb and Line

I n the last chapter we built and raised the walls. They're standing there, more or less straight. What's next?

Well, unfortunately, more or less straight isn't good enough. That's why you plumb and line. It's absolutely essential if you intend to do professional-quality work. And it's one of the few tasks in rough carpentry that requires the more patient mind-set of a finish carpenter.

Plumb and lining means temporarily bracing the walls with 2 x 4s to force them to stand "plumb" and "in line." *Plumb* (or level in a vertical sense) means the wall is standing straight up and down, not at a slant. *In line* means that you see a straight line when you sight down the long walls. You plumb and line with a level, your eye, occasionally a string, and now and then a rough terrain forklift (pettibone). When the framing is finished, you remove the temporary braces. If all goes well, the walls will stay plumb and in line.

But if you plumb the walls with a level that's out of adjustment, watch out! You'll be creating a nightmare that probably won't be discovered until much later — like when the finish carpenters try to install kitchen cabinets or hang doors. That's too late. True, you'll probably be off the job by then, unless you're working on a housing development. But don't think that lets you off the hook. You'll suddenly find that your reputation arrives before you do and no one will want you working on their jobs. Word of mouth is the most important advertisement you have in this business, so keep your nose clean.

Of course it isn't impossible to replumb and line a section of wall that has a floor bearing on it, is filled with plumbing pipes and electrical wires, and maybe even drywalled. It can be done. You'll be working on

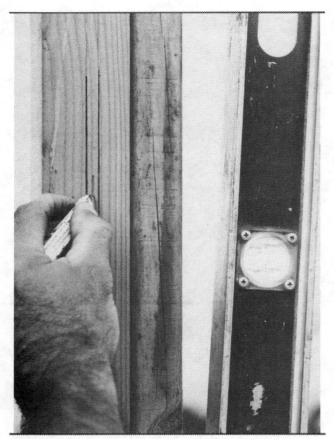

Figure 6-1   To adjust a level, first drive a nail part way in the bottom plate. Push the bottom of the level against that nail. Plumb up the level and make a mark on the wall. Turn the level so the other side is facing you. Center the bubble again and make another mark along the edge. Adjust your level to the middle point between these marks.

your clock so be assured that the general contractor will give you all the time you need to figure out a way. Always, *always, always* check your level before you plumb and line the walls.

Here's an easy trick for checking the accuracy of your level: Set a nail in the bottom plate of a wall, next to a stud. Put the bottom of your level against the nail. Then move the top half of the level until it reads plumb, and then draw a line on the stud along the level. See Figure 6-1. Now flip the level around so the side that was facing you is now against the stud. Center the bubble and draw another line. If the two lines are right on top of one another, your level is perfectly adjusted. If they aren't, then half the

distance between the two lines is where you want to be. Loosen up the screws that hold the bubbles, hold the level halfway between your marks, and retighten screws when the bubbles read level.

How to Plumb and Line Standard Walls

The first step is to attach the bottom plates firmly to the concrete slab or wood floor within the lines snapped by the detailer. If it's a wood floor, the wall builders should have nailed down the bottom plate approximately every 16 inches. On slabs, the exterior walls (and possibly some interior shear walls) have bolts sticking up through the bottom plates. Apply a nut and washer to each stud and torque them down. Use an electric impact wrench (like the one in Figure 6-2) if you have one. Otherwise a ratchet wrench works fine.

Use a small sledgehammer on the bottom plate to move a wall until it's on the snapped line. Make sure the wall is sitting in place on both ends. Also make sure that your channel marks (see Chapters 4 and 5) line up with the perpendicular wall lines before you tighten down the nuts.

Shoot the interior walls in place with a powder-actuated nail gun. Figure 6-3 shows the nail gun I like to use. It shoots hardened cement nails with a load about the size of a .22 caliber bullet. Drive nails about 32 inches apart. Be absolutely sure walls are where they belong before shooting them down. This step is known as *nutting and shooting*.

■ Lay Out the Braces

Now you'll need to spread out all the braces. First check with the foreman to see if there are areas that he wants kept free of braces. Sometimes you'll brace balloon or rake walls only up to a certain height for now, and then finish bracing them later.

Figure 6-2 Tightening bottom-plate nuts is a snap with an electric impact wrench.

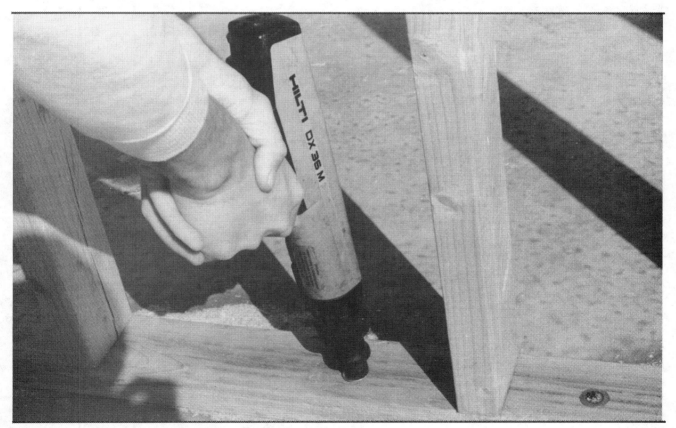

Figure 6-3 Shoot down interior walls with a powder-actuated nail gun.

Figure 6-4a   To hang a brace, first start two nails.

Figure 6-4b   Then hold the brace in place and nail it up. With tall walls (or short framers), you might need two people to do this efficiently. Have one person hold the brace while the other nails. Move throughout the house hanging all the braces at once.

Spread out all your bracing material before attaching any braces. That makes it easier to move about freely. Once the braces are up, trying to work quickly becomes just about impossible. Any wall that doesn't have a let-in brace will need a temporary brace to hold it plumb. Remember from Chapter 5 that a let-in brace is built into the wall by the wall framers. Where there's a let-in brace, you don't need to hang a temporary brace on that wall.

Sections of walls that span 14 feet or so without an intersecting wall running off of them need a line brace somewhere near the center. Walls that run perpendicular will be braced also, so they help hold the long wall straight. A line brace takes care of the section of wall that stands between the perpendicular walls.

Although the plumb braces are temporary, that doesn't mean they can be flimsy. Because they run about a 45-degree angle to the floor, they make good ladders. Tradesmen are bound to climb on the braces in order to get on top of the walls. Nail them well. Make sure all brace nails get a good solid bite. Nail the middle on plumb braces. The last thing you need is a wall that slips out of plumb because someone used the brace as a ladder.

There are no hard and fast rules on where to put the braces. Just be sure each wall is braced plumb. Start by spreading a brace for each wall that doesn't have a let-in brace. Use judgment when setting line

Figure 6-5 You may have to nail or shoot down a block into the floor to nail the line braces against.

braces. Sometimes you can work with 14 feet between braces. Some walls are so bent you'll need a brace every 4 feet.

The rules for brace layout are simple: Every wall needs to be braced *plumb* and *straight*, with enough strength to hold *steady* when you shake it by hand. Keep these three rules in mind and don't settle for less.

■ Hanging the Braces

When all the braces are laid out, start nailing them up. Begin by nailing the braces to the top plate or the top of a stud. Set two nails in the top end while the brace is still resting on the slab. Then lift it into place and nail it where it belongs (Figures 6-4a and

6-4b). If you're nailing to the top plate, make sure the brace doesn't extend above the plate. You don't want the braces in the way of the joists.

Then set two nails in the bottom of the brace, ready to drive into the bottom plate when the wall is plumb. You might have to cut down a brace to make it fit in a tight area. Nail or shoot down blocks into the floor if you need something to nail the line braces against. See Figure 6-5. But never shoot anything into a concrete floor (such as in a garage) that won't be covered later. That chips the slab — and is guaranteed to make your boss unhappy.

Figure 6-6 shows a typical room after it's braced off. There's a plumb brace hung with two nails into the top plate of each wall. Since the walls were relatively straight, we only put one line brace in the center of each wall. Again, two nails hold the

Figure 6-6   This is a typical set of temporary braces. Each wall has a plumb brace running against the studs from the top plate to the bottom. Since the walls were essentially straight, only one line brace was required per wall. These start in the middle of the wall and end at any convenient bottom plate.

top of the brace in place. With a line brace, you need to nail the bottom of the brace against the bottom plate of any wall within reach.

We used a crew of two framers to set braces for this home. First, we plumbed each corner of the room. Then, one framer sighted along the top plate of each wall. The second framer pushed or pulled the bottom end of the line brace until the middle of the wall was in line. When it was perfect, the framer doing the sighting would call out to nail it.

■ The Tools You Need

When plumbing a wall, use a level long enough so one end is on the top plate and the other end is on the bottom plate. Checking plumb against the

studs (as shown in Figure 6-7) doesn't work. Most studs have a slight crown that will cause the level to rock.

An 8-foot level works fine on 8-foot walls. For 9-foot and taller walls, either buy a telescoping level or build a jig from two short 1 x 4 blocks that you've nailed to a long, straight 2 x 4. See Figure 6-8. The 1 x 4 blocks hold the whole setup away from any bent studs or midspan blocks, assuring that you're only reading the top and bottom plates. Be sure to use a straight 2 x 4 for your jig. Then tape a freshly-adjusted level to the jig. You're in business!

The other tool essential for plumbing walls is a push stick. Some walls need to be racked in one direction or the other to make them plumb. Over the years, I've seen carpenters use many types of tools to wedge walls into alignment. Most probably work

Figure 6-7   Checking plumb against a stud only tells you how straight the stud is. Most studs have at least a ¼-inch bow in the middle of the stud. That creates a ½-inch margin for error at top or bottom when you hold an 8-foot level against the stud.

Figure 6-8   A jig made with a straight 2 x 4, two short pieces of 1 x 4, and some duct tape is the best way to check wall plumb. Notice how the 1 x 4 holds the jig away from the wall so midspan blocks don't create a problem.

fine. But I like my push stick the best. I use a piece of relatively knot-free lumber that's about 16 inches taller than the walls I'm plumbing. I place the board a little off parallel with the wall (Figure 6-9a) and wedge it into the pocket where the stud meets the top plate. I push up with my hands while holding the bottom end with my foot. See Figure 6-9b. Getting the wall to rack is usually a simple matter. If I go too far, I just tap the bottom of the push stick back a little to relieve some pressure.

Nine times out of ten this works fine. If the wall won't move, there's probably a temporary brace somewhere that's holding it in place. If removing that brace still won't ease the resistance, try a longer

push stick. Fit a 14-foot 2 x 4 into the angle at the top of the studs. Wedge the bottom end of the stick against a bottom plate of a perpendicular wall. Tap that bottom end toward the wall you're trying to rack. That should move just about anything. But be sure you haven't lifted the top plate from the studs before going on.

■ Now You Plumb and Line

Always begin plumbing walls at an outside corner of an exterior wall. Check that wall for plumb on both ends and in the middle if another wall intersects. If the bubble is offset to the right, your

Figure 6-9a   To make walls plumb, they usually have to be racked one direction or the other. I use a push stick made from a 2 x 4 that's about 12 to 16 inches taller than the wall. Wedge it in the pocket formed where the stud meets the top plate.

Figure 6-9b   Now throw your weight onto the push stick. The push stick slides down the stud as the wall racks away from you. Then move the bar back up the stud against the plate while pushing the bottom end of the bar forward. Hold the bottom end in place with your toe. If you don't move the bottom end, the wall won't stay racked. If you rack the wall too far, back the push stick off a little by tapping the bottom back.

wall has to move to the right. Usually the entire wall has to move a little either left or right. Try to get the wall within ³⁄₁₆ inch of plumb (for an 8-foot height) at both ends and in the middle. Sometimes you'll have to split the difference between two or three bubble readings along a wall. Under normal circumstances, it's OK to have one end ³⁄₁₆ inch too far to the left and the other end ³⁄₁₆ inch to the right. Any more than that and you might need to do some wall repair work.

Once in a while you'll have two wall ends that lean to the right and a wall middle that leans to the left. That usually spells trouble. Check to see if the bottom plate is where it belongs. If it's right, then

the top plate must be nailed wrong. Try a few measurements on nearby walls and see if you can find the problem. If you can't find it — or if it will take more than pulling a few nails to fix it — make a note on the floor so you can deal with it later.

If you're plumbing and aligning on a piece-work basis, problems due to bad plating or framing aren't your responsibility. Bring it to the attention of the foreman. Either the wall builders or the detail man messed something up. They should accept responsibility for the mistake. You certainly don't want it. Then again, it might pay you extra to fix

Figure 6-10   If you have interior walls tying into an exterior wall, just tack the braces when you plumb the wall. Adjust these braces as needed so you can sight a straight line down the top plate.

it! If you're working on a custom carpentry crew, complete the plumbing and aligning. Then go back and fix the problems.

If you're working with two people, have one person use the level while the other works the push stick and nails the brace when the wall is plumb. But sometimes you can't use a push stick. Sometimes your work space is too tight to fit a push stick. (Back in Figure 6-5, notice how small a wall they're working on. There's no room to get a push stick on a 45-degree angle.) And sometimes the wall just won't move. Plain old grunt work may be the only way to get some walls to move. Gather a few buddies and push with all you've got!

Continue plumbing walls around the exterior of the building until you're back to the starting point. Once the exterior is done, move to the interior walls.

Plumb them the same way. Once the wall is racked plumb, nail up either your temporary plumb brace or the let-in brace, whichever you have.

When you've completed plumbing the walls, check the sections in between perpendicular walls for line. You can do it alone with the help of a push stick, but it's easier with a crew of two. One person walks the top plate and sights the wall; the other nails the line brace on cue.

Here's a tip to make the job easier. Suppose you have a long exterior wall with a lot of interior walls running off of it. Just tack braces on the interior walls. Don't sink all the nails. Drive them just enough to hold the brace. When all the walls are plumb, jump up and sight the top plate of the wall (Figure 6-10). Does it look straight? Do a little adjusting on the braces, where needed. Then sink all the nails in.

Figure 6-11a   This is the wall before alignment.

Figure 6-11b   Afterwards, with all the braces in, it looks fine. Notice how the braces are kept high enough to give easy access down the hallway.

On a very long wall where sighting from one end to the other is too hard, use a string line. Run string from one end of the wall to the other, about 1½ inches inside or outside the top plates. Then measure from the string to the wall at several points. The distance should be 1½ inches at each point. Figure 6-11a is a *before* view and 6-11b is an *after* view of a wall straightened using string line.

It's important that long walls be straight, for two reasons. The first is aesthetics. It just *looks* better straight. The second is more practical. If they're not straight, nothing else will work quite right. None of the rafters, joists, drywall or roof sheathing will fit the way they should. You don't want every tradesman on the job blaming the framer for shoddy workmanship. Do it right and earn the respect of those who follow.

When you're finished, walk the building, shaking any wall that looks weak. Check all the bottom plates to be sure you've nailed the braces. Check walls at random for plumb. If this is your first house for a contractor, you can bet they'll check walls for plumb after you've gone home. Leave behind work that needs no excuses. That's the way to build a reputation that keeps you busy when others are starving.

Plumb and Line on a Balloon Wall

Very tall balloon walls can't be plumbed and aligned the way I've just described. These walls take more time and patience. Instead of a level,

Figure 6-12 Plumb and line with a rough-terrain forklift if you can. We'd broken a number of push sticks trying to persuade this wall to move. Twenty-foot tall 2 x 8 studs don't rack easily. One framer is on the wall, holding the plumb bob line against a block. The driver can jump out and check the bob as needed.

Figure 6-13 Hold the plumb bob line against a 2-by block flush against the top plate. When the tape on the ground reads 1½ inches, nail up the brace.

you'll need a plumb bob. Instead of a push stick, use a rough-terrain forklift or small crane. A 20-foot wall might not respond to your push stick.

Have someone on top of the wall hold a 2-by block flush against the top plate. See Figure 6-12. Drop a plumb bob from the edge of that 2-by. When the wall is plumb, the bob will be 1½ inches from the edge of the bottom plate. Push or pull the wall with the forklift until the bob is 1½ inches from the bottom plate. See Figure 6-13. Then nail up your temporary brace. This may seem like overkill. But on tall walls, there's no other way to do it. With a plumb bob you know it's perfect.

What Does It Pay?

If you start with accurate plating and decently-built walls, you can make good money in plumb and alignment work. Piecing out plumb and line will pay anywhere from 7 to 20 cents a square foot. On large projects, it's most efficient to use four people. One does the nutting and shooting, another is spreading braces, and two people are going through with the level to plumb and align.

Rolling the Joists

Architects design buildings with joists rolled up on walls, hung on beams and cantilevered for fireplaces. Joists are doubled up, headed out to avoid plumbing, laid out at 16 inches on center, and may even disappear completely if a truss roof is stacked! What does all this mean? A subject so thoroughly riveting comes around only once in every great book, so hurry, read on.

Joists are horizontal members that form the support for floors, ceilings, or a combination of both. On a one-story house, the ceiling joists (CJs) support only the drywall or panelized ceiling. With a two-story or higher project, ceiling joists for the ceiling below are also floor joists for the floor above. Because they support more weight, floor joists are usually larger (from 2 x 10 to 4 x 14) than ceiling joists (from 2 x 6 to 2 x 10).

Joisting, like wall building, is generally pretty straightforward. It only takes a day or so to learn the scope of the job. After that, it's a matter of picking up the little tricks to increase efficiency. The biggest problem most carpenters have with joisting is walking on top of the walls. There are grizzled veterans who have never been comfortable walking on a 3½-inch plate, 8 to 28 feet in the air. And then there are the ones who could go up there in a 75 mph wind and do handstands!

The trick to walking on a top plate is not to look down. Keep your eyes out ahead of you, and trust your feet; they know what they're doing. The survivors in this business have learned to respect their limits. That means not venturing out on a wall that just doesn't feel safe, whether it looks dangerous or not. And don't ever take risks just to impress someone.

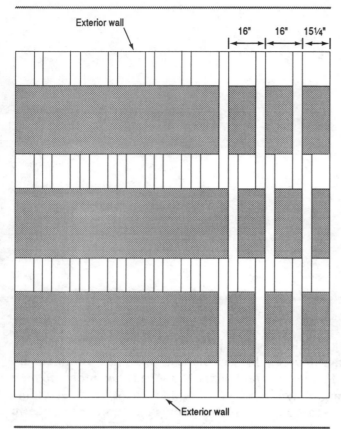

Figure 7-1 This is a floor plan that illustrates how to lay out joists that lap on interior walls. The top and bottom walls are opposite exterior walls. The middle two walls are interior walls. The first three joists are rolled up on the right. Notice how the middle package of joists are lapped *back* towards the wall that layout was pulled from. When you have to lap, it's better to lap back (creating an underspan) rather than lap forward (creating an overspan).

Planning the Joist Layout

In joisting, the first step is finding the direction the joists are designed to run. You can check the structural page of the plans or, if you're piecing, have the framing foreman show you. It's important to be absolutely sure which direction the joists go so you don't create an overspan. Joists are calculated to carry a maximum amount of weight. If you span them 2 feet longer than they were designed for, you'll create a weak floor whose structural members span beyond their limits.

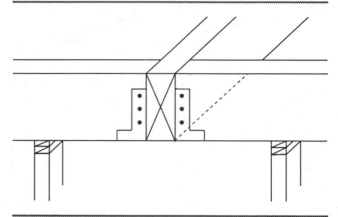

Figure 7-2 This is a typical *flush* beam. The bottom of the beam lines up with the bottom of the joists and the top of the walls, so the beam is buried in the ceiling.

A good illustration of this is the design of exterior decks. For their size, decks are designed with much stricter strength requirements than a standard bedroom floor. After all, dozens of people may crowd onto your deck to watch the fireworks on the 4th of July. More than one deck has collapsed due to this holiday!

If the joists don't span the entire width of the building, verify which interior wall the joists will lap on. Figure 7-1 shows how joists lap on interior walls. This is critical because not all interior walls are designed as bearing walls. Bearing walls are designed to carry a load, while nonbearing walls (partitions) only separate one room from another. Bearing walls must have thickened concrete footings. Without this footing, the normal 3-inch thick slab would crack and sink from the weight the finished floor and wall exert on it.

I worked on a tract once where all the joists had been lapped on the wrong interior wall. Since the problem wasn't discovered until the houses were finished, the framing contractor had to pay to have an interior footing dug and poured in every house. *Always be certain of the direction and the bearing points of joists before you begin framing.*

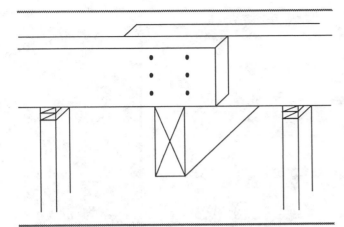

Figure 7-3 This is an exposed beam. The top of the beam lines up with the top of the walls.

■ Interior Floor Beams

Many houses have interior floor beams that have to be set before you can roll the joists. They're built into the floor to cut down on joist spans, to support stairs, or to create bearing points for loads from above. These are usually set by an hourly crew, but on some jobs the joist crew may do the job. The structural page of the plans calls out the floor beam sizes and direction as well as the hardware required to fasten the beam to the posts in the walls.

Some plans are more specific than others about whether the beams will be set flush or held low. A flush beam is set with the beam bottom flush with the ceiling line, as shown in Figure 7-2. When construction is finished, a flush beam is completely concealed. When a beam is set flush, joists are attached to the beam with hangers. The majority of floor beams are set flush, with the joists hanging on them.

When a beam is held low, the top of the beam is in line with the top of the wall. Figure 7-3 shows how the entire beam is exposed below the ceiling line. Joists lap over the top of the beam.

If the beam is held low, it's probably an exposed resawn or rough-textured beam. Always check with the supervisor, general contractor, or owner concerning the placement of beams *before* you do any cutting. Aesthetics as well as head clearance are two major factors with floor beams.

If the house you're working on has a number of interior beams, it's always a good idea to use a crane or pettibone to lift them in place, as in Figure 7-4. This is a lot safer — not to mention easier on your back. Some people argue that they can get the beams up cheaper with muscle power. That may be true. Just don't expect your crew to have much energy left after they've spent the morning lifting beams.

Preparation is the key to using a crane efficiently. Get *all* your beams cut and dressed with hardware on the ground. If you're setting the beams, it's a good idea to put the hangers on beforehand. Determine your layout and transfer it to the beam. It's a lot easier to do it on the ground than from a ladder. Look at Figure 7-5.

Figure 7-4 If you have the luxury of a pettibone on the job, all the better. They make easy work out of installing beams.

Figure 7-5 It took a lot of preplanning to establish the layout for the 2 x 12 ceiling joists that will hang from these beams. The house had exposed 6 x 8 corbelled fake rafter tails applied after the 2 x 10 rafters were stacked. It wouldn't be fun to put these hangers on once the beam was set.

Figure 7-6 Interior beams should be cut and labeled long before your crane shows up to install them. If any hardware is required, it's best to install it to the beams on the ground before they go up.

Figure 7-6 shows beams ready for the crane to lift. If you have a lot of lifts of wood that need to be taken up high, renting a pettibone is far more economical and handy. Always have plenty of hands around when you're putting out good money for a crane. Figure 7-7 shows how two people are needed to set beams with a crane.

■ Double Joists and Head-outs

Typically, you'll have to place a double joist under any upstairs wall that runs parallel with the joists, and on both ends of any cantilever. See Figure 7-8. When I'm going over the joist detail, either on the plans or with the framing foreman, I like to write all this information down on the concrete with a keel wherever it occurs. Later, when I'm up on the walls, all I have to do is look down to see exactly the numbers I need.

Figure 7-7 A lot of preplanning and a couple of extra hands are essential for a smooth day of beam setting.

Once you've established where any doublers are located, make a note, like "18'6" outside" directly below where the doubled joists go. Then, when you're laying out the joists on the top plates of the walls, you'll know to make a layout mark, 18'6" from the outside of the building to a doubler. If you're doing the joists on a piecework basis, the framing foreman will probably make notes on the slab for you to reduce the chance of disagreements later on.

Don't put a joist where plumbing drains will be located. See Figure 7-9. The plumbers need space to install drains for toilets, bathtubs and showers. The plans will show where the plumbing fixture drains are to be located. It's the framer's responsibility to avoid placing any wood joist that would conflict with these drains. It's up to you to scale off of the plans to find where to put your head-out.

Look closely at the floor plan of your building and you'll see toilets and showers drawn in place. Now scale off any interior or exterior wall to locate

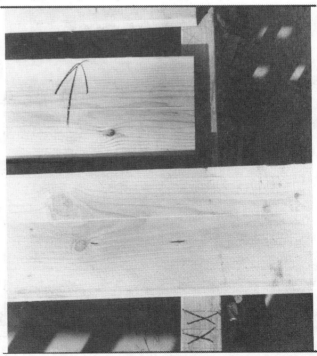

Figure 7-8 Identify single joists by a single X and doublers with XX as pictured. The doubler shown is the left side of a cantilever. The plan called for doublers on both sides of this cantilever.

Figure 7-9 You'll have to build oversized bays in the floor to allow the plumbers room to install their drains. These are known as head-outs. Scale off the plans where the toilets, showers, and bathtubs are drawn in to locate approximately where these head-outs will be needed. Sometimes the joist layout will allow you to center a bay where the drains are located, so you won't have to provide a head-out. The 2 x 4 that runs under this drain was installed by the plumber, apparently as a support.

where the drain is drawn in. When you locate (and double-check) the location of the drain, use your keel to mark the center of the head-out on the slab. See Figures 7-10a and b.

In most cases, if you head-out one joist and match that width in the other direction, you'll leave the plumber ample room. In other words, when you create a head-out, you're removing the section of joist that would interfere with the drain by "boxing out" perpendicular to the direction the joists are running. Figure 7-10b shows how the head-out supports the remaining length of joist. If you removed the entire joist, the layout would be 32 inches on center, which isn't allowed in most cases.

■ Cantilevers

Note any cantilevers on the plans and discuss them with the foreman. Cantilevered sections of joists provide a little extra floor space beyond the perimeter of the wall below. See Figure 7-11. Closets, bay windows, fireplaces, and small nooks can cantilever from a room without robbing it of any floor space. Upstairs decks are commonly designed

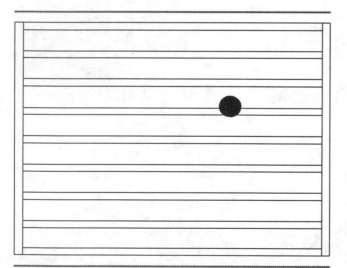

Figure 7-10a Here's a typical section of joists rolled up at 16 inches on center and ready for the head-out to be built in. The circle represents the exact location of the drain as shown on the plans.

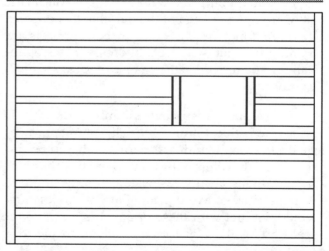

Figure 7-10b The joists are cut and the two head-outs are nailed in place. The two side joists are now shown as doublers. Inspectors often require the two joists that support the head-out to be doubled up.

to cantilever out past the downstairs exterior wall. By cantilevering the joists, you avoid having to install a supporting beam underneath the joists. The cantilevered joists are usually ripped down so they slope away from the building ¼ inch for each foot of length. This assures good water drainage from the deck.

■ Laying Out the Joists

When you've finished marking notes on the slab, it's time to lay out where each joist will rest. You'll mark these joist positions on the top plate of each end wall. This is the most critical step of joisting. Use an X on the top plate to identify a single joist, and XX for a double joist.

Since floor sheathing comes in 8-foot lengths, the center of the sixth joist must be exactly 8 feet from the far edge of the first joist (assuming 16-inch center-to-center spacing). Shift the layout for the first joist back ¾ inch. Then lay out the remaining joists at 16 inches on center. The edge of each panel should fall on the center of a joist.

Start on the outside corner of the building. Hook the tape and measure in 15¼ inches for the first joist. Set a nail on this mark so you can hook your tape on it like in Figure 7-12. Continue down the length of the building, making a mark every 16 inches to represent "to" each joist. When you come across a doubler or cantilever note on the floor, lay it out on the top plate as well. Notice the single and double layout marks back in Figure 7-8. Now, if the joists span the entire width of the building, go to the opposite parallel exterior wall and lay it out exactly the same way.

If the joists lap on an interior wall, you'll also have to lay that out. When you lay out the lap wall, remember to specify which way the second section

Figure 7-11 When you roll the cantilever joists, block the common bays as you go. Then you can cut the special blocks when you're ready to do all the rest of your cutting.

Figure 7-12   Lay out the first joist at 15¼ inches, and then tack a nail in the top plate at this layout. Now hook your tape on the nail and lay out the remaining joists at 16 inches on center.

of joists will lap. In most cases it's best to lap back, creating an underspanned first bay, rather than lapping forward and creating an overspanned first bay. Look back to Figure 7-1.

When the ends of two joists meet on an interior wall, rarely do the ends butt. Instead, they lap side by side. If you lay out the first joist at 15¼ inches from the outside wall, to the joist, then the second lapping joist will have to go either 15¼ and back, or 16¾ to it. It's better to go 15¼ and back.

If you have an even number of joist packages, remember that you'll have to compensate for the direction of lap when laying out the other exterior wall. This doesn't apply to the layout in Figure 7-1 because there are three packages of joists, so the layout on opposite exterior walls would be the

same. On the front wall, the first joist will be 15¼ inches to it. The lap wall will be 15¼ inches to the middle of the lap. And on the back wall, the joist will be 15¼ inches and back.

Spreading the Joists

When joists are all laid out, start spreading the joists as in Figure 7-13. Take the time to crown them. Sight down one edge of each board and you'll usually find a slight curve in one direction. This is the crown. The crown should be up when the joist is installed. I like to crown the joists on the ground because it's easier there than at ceiling level. As you crown each joist, mark an arrow showing the direc-

Figure 7-13 Most of the labor involved in joisting is hustling the joists up on the walls and spreading them in place. Having a crane or pettibone lift the joist packages is a real backsaver. Piece workers are generally compensated for loading their own wood.

tion of the crown. When you spread the joists in position on top of the plate, keep all crowns facing the same direction.

If you're spreading the joists in a section that requires a flush beam, set the joists in the hangers as you spread them. Nail the joists as you spread them like the framers are doing in Figures 7-14 and 7-15.

Once joists are spread out on top of the plates, lay out lumber along the outside edges for rim joists. Figure 7-16 shows crowned joists (arrows pointing to the upper edge) and the piece to be used for the rim joist.

Begin at one of the outside corners of the building. As you roll the joists (tip them up into place), make certain that all the crowns are up. Roll up the

first joist and the rim joist and pound three nails through the rim into the joist. Toenail down through the rim into the top plate every 16 inches. Be sure the rim joist stays exactly flush with the outside of the building.

On shear walls, it's not enough to simply nail the rim joist to the top plate every 16 inches. That wouldn't make a sturdy connection between the wall and the floor. Instead, use A35 metal straps to connect the wall and the rim joist. Look at Figure 7-17. The shear wall schedule in the plans should identify the correct spacing for straps.

Toenail the first joist flush to the outside of the wall that it bears on. Then roll up the next joist. When the bottom of the joist is on the layout mark and the joist is standing straight, nail it to the rim. Drive a toenail on either side of the joist down into

Figure 7-14 Having a plank to walk on when spreading joists up high helps even the most sure-footed. The two center walls that run under these joists will need to be toenailed into the bottom of the joists once they're all rolled. This will hold it straight once the temporary line braces are removed.

Figure 7-15 Fill the holes in hangers with 1½-inch hanger nails, and then toenail the bottom of the joists into the center wall.

Figure 7-16   Spreading all the joists with the crowns pointing in the same direction will help when you go to roll them. Spread rim stock where it will be needed. Notice how the joists were all kept back from the edge of the wall when they were spread to allow for the rim to be rolled up. The first joist in this package was rolled up previously and used as a rim for the run of joists in the background that run perpendicular to it.

the top plate. See Figure 7-18. To save time, always work in one direction, from one side of the building, then turn and toenail back to the other side.

Cantilevered joists have to be blocked where they overhang the supporting wall. Lay down joist blocks for any cantilever area at the same time you lay down your midspan blocks. See Figure 7-19. I like to nail in the blocks as I roll up each joist. If you roll all the joists up first, drive toenails into the top plate, and then come back to set blocks, it can get a little tight. Instead, block as you roll. The same

Figure 7-17   On shear walls, nailing the rim to the top plate every 16 inches doesn't create enough of a structural connection from the wall to the floor. These walls need A35s connecting the two. The shear wall schedule will tell you the spacing.

holds true for laps. Nail the lapped joists together, but don't nail them into the top plate until you slide a block in place.

Notice in Figure 7-11 that the rim joist in the foreground stops right where the cantilever doubler begins (the only cantilevered joist rolled up). This was the point where I began rolling up the joists. By starting my rim joist tight against the first cantilever joist, I was able to let the rim run wild past the outside of the building and avoid having to measure the rim. After the cantilever joists are all rolled, you can measure and cut the last little piece of rim joist. So you could run all the rim for this section and only reach for your tape once. It's tricks like this that make you the money.

Now you can roll all the normal blocks on the cantilever. There might be a few special-cut blocks that need measuring, but normally the cantilever joists will be on layout with the rest of the joists and have normal 14½-inch blocks. The two outside cantilever joists are probably not on layout and will require special blocks. And any rim you ran wild will need to be cut flush with the first joist, which is nailed flush with the outside wall (Figure 7-18).

Figure 7-18   After getting three or four nails through the rim into the joist, put a toenail down into the top plate of the wall. When you've finished rolling this run of joists, turn around and get another toenail down into the top plate on the opposite side of the joist. This is commonly known as *back-nailing*. In this situation the rim was run long and will be cut off flush with the outside of the building as soon as another joist is chased down.

Figure 7-19   Midspan blocks go every 8 feet of joist. Snap a line, and then set three nails in each bay, staggering the blocks either side of the line from bay to bay. I would caution against holding the block in the manner shown in the photograph. First, the bouncing of the joist against the block could award this finger a blue-ribbon blood blister. Second, if he happens to aim a little high, his finger's in line for a 24-ounce waffle sandwich.

The only thing left to do in this section is to cut the ends of the cantilevered joists and hang a rim on them. The plans will give you the measurement for the overall distance that the cantilever extends beyond the exterior wall.

Remember, do as much framing as possible before cutting. Letting the rims run wild lets you keep the tape in your pouch. It's easy to cut rim joists by sight once they're installed.

If you're lapping the joists over a supporting interior wall, install rim joists on both opposing exterior walls and let the excess run wild past each other over the interior wall. When you've installed blocks between each joist at the laps, cut off the excess material as close to the lap wall as the saw will comfortably go.

You'll need rows of midspan blocks every 8 feet of joist that spans from one supporting wall to the next. If you're piecing the job, the framing foreman should supply common blocks 14½ inches long. You'll have to cut any special blocks. Snap a line across the joists where you need blocks.

Figure 7-20   Any lower wall that runs parallel with the joists will need to have ceiling backing installed across its top.

Figure 7-21   Have one person drop down low to cut and hang all the backing. The backing has been installed in the upper right corner of the photo. You can see the next piece hanging on the joist in the middle of the picture. If you have the material, you can use a 2 x 6 to cover both sides on the top of a 2 x 4 wall. Otherwise you have to cut and install two 2 x 4s to back out the ceilings on both sides of this wall.

Tack in place all the nails for these joists, staggering them left and right of center from one bay to the next. You can see the nails in Figure 7-19. Because these blocks don't need to be in an exact straight line, the blocks will stagger from one side of the snapped line to the other. This makes nailing them a lot easier.

Slip a block down between the joists and drive the nails you tacked in place. That saves fumbling for nails. Just bang them right in. Get all the blocks nailed on one side. Then turn around and go back across the joists, nailing the other side. Make sure the blocks are all flush with the top of the joists.

There are two advantages to nailing all the blocks in one direction and then the other. First, it's safer. Turning one way and then another for each joist risks a fall. Second, it saves time because you take fewer steps.

Where interior walls run perpendicular under the line of joists, drive two toenails down through each joist into the top plate. This holds the wall plumb once the floor is sheeted and temporary braces are removed. Check to make sure any joists that were set in hangers are completely nailed off.

Placing the Ceiling Backing

Figure 7-20 shows the last step in joisting. Any wall that runs parallel to the joists needs backing as a nail base for drywall on the ceiling. Figure 7-21

Figure 7-22   Ceiling joists are typically smaller than floor joists because they only support the drywall that's hung on them. They also require no rim, as the rafters will be nailed up against them later.

shows this backing as seen from below. You'll probably want to measure and cut all your backing while standing on the floor below. By setting a temporary nail in one edge, you can hang it from below on a joist. Then go back up on the joists, grab the backing, and nail it in place. Make sure it's nailed down securely. Remember, the drywaller will be pounding up into the backing when he nails up the ceiling drywall. If it's not secure, the finished ceiling will bow up at the edges.

Be sure the bottom of the backing is level with the bottom of the joists. If the backing sits too low (either because of a big crown in the joists or because the slab is uneven), you'll have to install *float backing* between the joists. First, cut some 2 x 4 blocks 14½ inches long. Nail these blocks between the joists so the bottom of the blocks are 1½ inches above the bottom of the joist. Then, when

the backing is installed to the blocks, it will end up flush with the bottom of the joists. This float backing will be a little above the parallel wall. But toenail it down into the top of the wall anyway. Just don't overdrive the nails — you don't want to pull the backing down.

Ceiling Joists

Ceiling joists are rolled in the same way as floor joists — with a few exceptions. First, ceiling joists don't require a rim joist along the exterior wall. Look closely at Figure 7-22. Because the roof rafters will rest on top of the wall alongside each ceiling joist, a rim joist would be in the way.

Figure 7-23 A *catwalk* is nailed to the top of ceiling joists to tie the structure together, to help keep the joists in the same plane so that the drywall ceiling is straight, and to keep the middle of the joist on layout.

Figure 7-24 These Silent Floor truss joists have a laminated top and bottom chord, pressed onto a piece of plywood. They're lighter than conventional joists, and can span greater distances.

Once the rafters are rolled (or set in place), insert a block in the space between each joist. That strengthens the whole setup, much like the blocks did the cantilever section. Ceiling joists have two main functions. First, by spanning between the bases of two opposing rafters, they add horizontal strength for the roof. Second, they support the drywall that makes up the ceiling. Because they only support the ceiling drywall (as opposed to floor joists that support the drywall plus the weight of whatever is resting or moving about on the floor) they can be much smaller than floor joists. And they don't need midspan blocks every 8 feet of joist length like floor joists do. However, they do need to be blocked at every lap joint to insure the lateral (horizontal) strength that the roof depends on.

In place of the midspan blocks, nail a board (called a catwalk) on top and perpendicular to each line of joists, as shown in Figure 7-23. The catwalk keeps the joists from twisting as they dry, and also holds any slightly-bent joists exactly on layout. When you install the catwalk, nail it on top of the first ceiling joist. Then mark off a 16-inch on center layout along the catwalk. As you nail the catwalk into the top of each joist along the layout, make sure it lines up with the layout mark you made on the catwalk.

Truss Joists

A lot of builders are using truss joists these days. Figure 7-24 shows a tract I worked on recently where we used Silent Floor truss joists. No midspan blocks are needed with truss joists. That makes the floor quieter because midspan blocks squeak if they're cut too short. In Figure 7-25 you can see

Figure 7-25   Flush beams are built into the floor with the joists hanging off of them. With truss joists, your beams will likely be *Micro-lams*, which are fabricated somewhat like glu-lam beams.

flush beams with Micro-lams hanging off of them. Micro-lams are fabricated beams that are made much the same way that plywood is, with layers of wood grain opposing each other.

Premium-quality lumber is cut from old growth forests. Wood like that is both more expensive and getting scarce. That's why you'll see more and more manufactured lumber like Micro-lams and truss joists on jobs in the future.

Figure 7-26 shows the structure of a typical commercial ceiling. A glu-lam beam supports truss joists laid out 8 feet apart. The truss joists support prefabricated panels made of 2 x 4s and plywood. This design allows for an open floor space with a minimum of interior walls. Some type of decorative ceiling will be installed later to hide the structure.

Figure 7-26   The main support for this commercial ceiling is provided by the massive glu-lam that runs down the center of the room. The truss joists are laid out 8 feet on center, then prefabricated panels made of 2 x 4s and plywood are placed on top of them.

Joists on Concrete Stem Walls

Joisting on stem walls is very similar to joisting on the tops of walls. But it's a lot easier than climbing up and down a ladder, and a whole lot safer! Everything we learned about layout, head-outs, cantilevers and blocking applies here also. Before you can roll any joists, though, you need to apply a pressure-treated 2 x 6 on top of the stem wall, all the way around the perimeter of the building. Bolt the 2 x 6 down to the stem wall with the anchor bolts that are installed in the concrete when it's poured. Notice the 2 x 6 on the stem walls in Figure 7-27.

Once the 2 x 6 is down, proceed with business as usual. Mark the layout for the joists on the 2 x 6 as before. Then it's easy to spread the joists, put on the rim and roll the joists (Figure 7-28).

Exposed Beams

Exterior beams are often exposed, so every cut needs to be as tight as possible. This brings into question your technique for measuring and establishing odd angles. Although most jobs aren't nearly as cut up as the one pictured in Figure 7-29, it serves as a good example for the finer points of beam cutting.

Figure 7-27 Some houses are designed with a wood floor resting on concrete stem walls for the first floor, rather than a concrete slab. With these designs you'll need to first bolt down a 2 x 6 green plate around the complete perimeter stem wall. Girder beams are set perpendicular to the joist across the middle of the floor to cut the joist's span in half. The short wall in the middle is a bearing point that carries all the way up to the roof.

Figure 7-28   Then you'll spread and roll the joists as usual.

Figure 7-29   With a beam layout as cut up as this, it's best to make templates out of 2 x 6s to find the length and angle of each individual beam. This helps to keep the cuts nice and pretty. With red iron supports, you have a difficult time finding the angles. Don't trust the plans! The actual placement of the iron is what you're concerned with.

If your project requires precision cuts, or your beam supports offer no easy way to establish the angles, you might want to use temporary templates to find the exact length and angle of each beam. If your beams are 4-by wide, use 2 x 4s as templates so you have the correct width. If they're 6-by wide, use 2 x 6s. Use boards that are longer than you need so they overlap at the angle cuts. That makes it easier to establish the exact angles at each union.

When you have all the templates in place, take one down and transfer the length and angles directly onto your beam while it's still on the ground. Cut the first beam and hoist it in place. Now check your templates on either end of the beam while it's in place, as in Figure 7-30. Are the angles still true? If not, adjust the template before you cut the next beam in line. That way you can have tight cuts all the way around. Many times the supporting members aren't raised in the exact perspective angles that the plans call out.

Figure 7-30 With one beam cut and in place, you can check how it fits with a template on both sides. Make adjustments on your templates if needed before you cut the next beam.

Figure 7-31 Once all the beams are in place, you'll end up with a tight fit all the way around. The angles at each intersection of this setup changed 7 to 10 degrees, but using templates we were able to get every cut reasonably tight.

By using templates, you can establish the angles and end up with tight cuts. Figure 7-31 shows all the beams cut and in place.

And What Does It Pay?

One carpenter working alone can install joists, but it's a lot easier if you have a helper. Either way, you can make good money once you know the tricks that make the job go faster. Simple joisting, where there are no beams to cut around, and no laps or cantilevers, usually pays somewhere around 2 to 4 cents over the joist size, per square foot of floor covered. To put it another way, 2 x 12 ceiling joists covering 2,000 square feet of floor would pay 14 cents a square foot, or $280. In most cases, though, the designs are cut up and it will pay from 20 to 50 cents, depending on how cut up it is. Beam work usually pays by the hour, or at an agreed-on price.

EIGHT

Subfloor Sheathing

After the floor joists have been installed, they're covered with the subfloor sheathing. The term subfloor refers to the structural, rough sheathing, as opposed to the finish floor. The finish floor is the nonstructural component: hardwood strip flooring, marble, terrazzo, tile, carpet, or sheet vinyl that's installed during the finish phase of construction. Fifty years ago the most common subfloor material was tongue and groove 1 x 6 boards, laid diagonally across the joists. Installing diagonal floor sheathing was slow work. The 1 x 6 was usually high-grade pine, fir, or even redwood. Diagonal sheathing made a sturdy subfloor and is still used occasionally today as a finish floor laid over a plywood subfloor.

Laying a plywood subfloor is much faster and more economical. Today, most floors are sheathed with 5/8-, 3/4- or 1 1/8-inch thick tongue and groove plywood, which comes in 4 x 8-foot sheets. Tongue and groove plywood (T&G) is faster to install, and in most cases, it's every bit as strong as the 1 x 6 floors of yesterday.

Sheathing Materials

The T&G plywood used for floor sheathing is usually 3/4 or 5/8 inch thick. On one side of the sheet, the knots are plugged with wood dough and sanded down smooth. This side goes up. T&G plywood is manufactured primarily for subfloor use, so any T&G you come across should be a high enough grade to use as a subfloor. The structural floor sheet on your plans should call out the thickness needed for the particular house you're working on.

Figure 8-1   Lay beads of glue in 4-foot strips down the whole line of joists you plan on sheathing.

There's a less-expensive type of T&G sheathing that resembles particleboard in its construction and appearance. Its official name is oriented strand board (or OSB). One side of it is stamped "This side up." Unless you plan on covering the subfloor with lightweight concrete, or you have wood underlayment covering your subfloor, I would avoid this material. If your joists are in the least bit overspanned the floor will feel very spongy and weak. If the sheathing happens to get moisture on it during construction, the problem is even worse. It also has only about half the sound-deadening properties of regular plywood. That's an important consideration in multi-story apartments and condos.

I'm a strong advocate of new materials that will conserve our precious timber resources. OSB has its place in construction, of course — but I don't

think that place is as a subfloor. But if you do decide to use it, try extra hard to keep moisture off it and use the ¾ inch thickness. It works far better than the ⅝ inch.

There's one exception. That's when sheathing is covered with 1½ inches of Elastaseal (a lightweight concrete) for fire and soundproofing between floors, or as a permanent bed for applying marble, terrazzo or tile. Under lightweight concrete, OSB does a great job, making the savings worth it. Whether the subfloor will be covered with concrete or wood underlayment, you lay all T&G subfloor the same way.

Sheathing Techniques

Do a little planning before you lay the first row of sheets. Find the longest, least cut-up area. Start there and work away from it, towards the more cut-up areas.

You always lay plywood subfloor perpendicular to the supporting members. In other words, the grain you see on the sheet of plywood (which always runs lengthwise) runs perpendicular to the joists. With ¾- and ⅝-inch plywood, the joists should be no more than 16 inches on center. With 1⅛-inch plywood, joists can be 24 inches on center and the floor will still be solid and rigid under normal loads. Of course, joists have to be sized correctly for the span. Plywood floor sheathing doesn't add any strength to the floor structure. No plywood, however thick, will help a joist that's overspanned. To avoid a springy floor, you must use joists of the correct size and species for the span.

Start by snapping a parallel line 48 inches from the outside rim joist and perpendicular to the main body of joists. Spread out as many sheets as you can in roughly the correct position. But leave the first 48 inches of width clear. The tongue of each sheet of plywood should point the same direction, toward the starting point. The grooved edge should be facing the direction where the next panel will be installed. When you drive each sheet

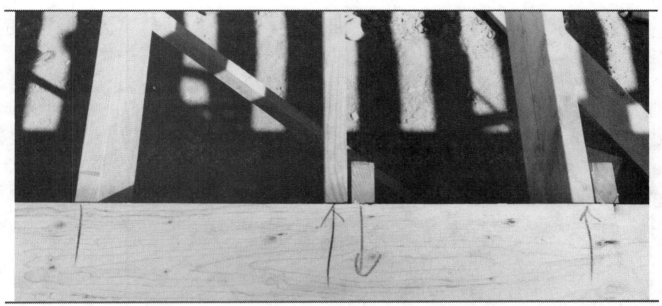

Figure 8-2   Before you start on the next row, go down the line and mark out where the joists pass under each sheet. That saves a lot of time when you nail off the floor. This photo shows where a lap occurs. The arrows show the position of the joists, and the direction the joists are lapping.

snug by smacking your jig against it, the grooved edge is more able to stand the blows you'll be dealing it. That saves the tongue edge, the weaker of the two edges.

In the first 48-inch width, spread a bead of adhesive along the rim and on each individual joist. Look at Figure 8-1 (except he's spreading the glue on the second 48-inch width while standing on the first). Place a sheet down in that 48-inch space. Make sure the tongue faces toward the wall where you begin. Place this first panel exactly straight on the 48-inch line. Always follow the perfectly straight line you snapped. If you have to choose between keeping the sheet on the line or keeping it flush along the outside of the building, always go with the snapped line. If your building has a severely bowed exterior wall (which the rim would follow!), you might have to add backing in the floor to support the plywood where it leaves the rim. In any case, stay with your line!

On your first piece, make sure the leading edge falls exactly over the middle of the last joist. For a 16-inch on center layout, that would be joist number six, not counting the rim joist you started on. When it looks good, nail it down. If the joists weren't laid out accurately, you'll have to cut the first sheet so it splits the last joist. Continue until the first row is down and nailed with at least one nail in each corner and into each joist that passes under the sheet.

The first row of sheets needs to be tacked down pretty securely to keep it from moving when you start banging in the next row. Use a pencil or keel to mark on the first row of sheets where the center of each joist passes below. See Figure 8-2. When you're over a lap, like the middle joist in the photo, use arrows to show the change. When you go to nail off the floor later, this little change can be a real mystery if it isn't labeled properly. You'll have to know exactly where these joists fall when you nail off the subfloor.

Start the second row of panels with a 4-foot by 4-foot sheet of plywood so your end joints are staggered. Begin the second row by spreading out the glue and placing the first sheet down with the tongue leading into the groove of the first row. The trick now is to drive the tongue of the second sheet fully into the groove of the first.

Figure 8-3 You might try using a double jack (or large sledgehammer) to persuade the sheets together. Be sure to lay a board down to protect the groove when you're hammering. You'll surely waste the groove if you don't.

■ Fitting the Tongues and Grooves

If the tongue or groove has been damaged, do a little repair work before smacking them together. Every tongue should be fully seated in the groove. On really tough sheets, try driving them together with a double jack (or large sledgehammer) against a 2 x 4 laid flat along the groove. See Figure 8-3. At the same time, apply pressure with your feet at the joint to guide them together. Long legs and arms come in handy for work like this.

Seat the tongue fully in the groove of every adjacent panel. If it's not as tight as possible, you'll create a kink in the line of sheets. Since joints are always staggered on adjacent rows, any panel that

isn't fully seated will affect two panels on the next row. Any irregularity will tend to get worse, not better, as you lay more sheets. There are three easy ways to avoid these problems.

■ First, always make a quick check of each tongue and groove before you attempt to join them together. The trip between the factory and the job site usually wreaks havoc on a few of the sheets. Clean out the grooves and repair any tweaked tongues to make a snug fit.

■ Second, you could have an assistant guide the joints together as you drive them snug with a sledgehammer. That's fine if someone is paying you both good money to lay subfloor. But on a tract job, two people can't

Figure 8-4   With a jig like this, you can sheath floors alone. Adjust the pressure on the T&G joint with your toes to align the tongue into the groove.

survive on a nickle per square foot of floor. If you plan on making any money piecing out floors, a one-man sheeting crew is the only practical method.

▌ There's a third choice. You don't need an assistant if you use a jig that provides enough power and maneuverability to persuade the pieces together. Resourceful framers have invented any number of quickly-built devices to help join the tongue and groove. Try the quick and dirty tool shown in Figure 8-4 that you can make from scraps lying around the job, or make one you like better. With a couple of good slams, the device in Figure 8-4 will work wonders. Control the T&G joint with your toes, applying pressure where needed, while pulling the sheet snug with the jig. This is the way most piece

sheeters work on small jobs. On a large job, they might use a laborer to lay down glue, spread the sheets, and then help guide the pieces together when needed.

Up to now, we've assumed that all the joists run in the same direction. If the joists change direction (as in Figure 8-5), you have to change panel layout with them. Cut the panel to fit over the middle of the last joist. The other half of that joist has to serve as backing for the row of panels that run perpendicular to it.

When you roughly lay your sheets in their final position, use the method demonstrated in Figure 8-5. Hold the piece along its grooved edge and maneuver the plywood so that the two edges will land half on their perspective joists. As you drop the sheet, control its fall with one of your feet. As it

Figure 8-5   Drop the sheet in place endwise. It should fall directly in place so you don't smear the glue, rendering it useless. Try dragging your foot down the face of the sheet as you drop it. As it falls, pull the sheet back toward you with the sole of your shoe. Notice that the joists change direction in the middle of this photo.

falls, drag your foot back towards you. This will cause the tongue to suck into the grooved piece that's already down. This is the cleanest and fastest way to lay sheets.

Any other method, including using two people to place the sheet, ends up smearing the glue off the joist and rarely gets the tongue started in the groove. Also, if you drop it alone as shown, you never leave the previously decked area. When setting it in place with another person, you have to venture out onto the joists and risk a fall. The only time I ever saw stars was from a fall I took between joists. It put me in bed for a couple of days and lost me the work. Try to learn this safe and money-making trick.

Make sure each sheet of plywood is firm and snug before moving on to the next sheet. Keep the rows straight. If you start losing a straight line, it can only get worse as you add more sheets.

■ Notching Around Drains and Pipes

If the plumbers have installed their drains and copper lines before you sheet, you'll need to notch around them as you go. See Figure 8-6. Try to keep your cuts as tight as possible, but at the same time don't spend all day making pretty cuts. Most of your cuts will be buried under the bottom plates of the second story walls, anyway.

Figure 8-6   On large jobs, the plumbers will more than likely begin work before sheathing is laid. If so, you'll have to notch around all the pipes.

When making your cuts, there are only two things to remember. First, you'll have to make the cuts a little big in the direction your sheet is moving when you set the tongue into the groove. If you make it too tight, the sheet won't be able to slide into place. As you hit it with your jig, you may move the pipes out of line (or worse, damage them). Always leave a little room to work with.

Second, always make sure that you're measuring off the right edge. When you measure towards the pipe you're notching for, take your reading off the *finish edge* of the installed plywood. Place your tape against the groove, and take your measurement. Then when you transfer to the sheet you're cutting, hook the sheet above the tongue and mark out your numbers. Don't hook the tongue or you'll be adding ⅜ inch. Remember, when the sheets are

together, the grooved edge sits tight against the upper and lower finish edges of the sheet with the tongue.

■ Nailing the Sheathing

Once you've laid the sheets, trimmed the edges and marked the joist locations, get out your nail gun. You're finished with the hard work; it's time to make the easy money.

Most floors are nailed with pneumatic nail guns. The nails come in strips held together with plastic, sold in boxes of 5000. Take care any time you use a nail gun. Always wear some sort of eye protection. The plastic that holds the nails together bursts off the strip as each nail is driven. Flying plastic can sting your eye — if you're lucky. If you're not, it can blind you.

On most floors, you'll use 8d nails spaced 6 inches on the edges and 12 inches in the field. Check the plans or ask the supervisor what nails and spacing are required. Then fire up your compressor and test fire a couple of clips to make sure the compressor is adjusted to keep up with you. Drive the nails just deep enough so that the head of each nail is flush with the plywood. Be careful not to go too deep. Of course, nailing should press the sheet tight against the joist.

With your gun set for the correct nail depth, concentrate on nail spacing. You don't want to come back later to fill in extra nails. Listen as you nail. When the nail hits a joist, it makes a solid sound and the sheet sucks down tight. If you miss, it's more of a flat, hollow sound, and the plywood lies dead or bounces. Also, when you hit a joist, the gun bounces back just a little; on a miss it doesn't. Send in another nail next to your miss and listen again.

Edge nailing is easy. You know where each joist is. Nailing in the middle of each panel can be harder. You have to estimate the joist position from the rough marks you made on the top of each panel. Listen carefully and you'll hear which nails are hits and which are misses. Trust your ears. With a little practice, you'll be nailing along with the best of them.

Most squeaks in floors are due to faulty nailing. If your nail gun doesn't have enough pressure, the nails won't hold the plywood down tight against the joist. Even with the heads buried in the top ply, they might not be pulling the sheet down tight. When someone walks near this nail, friction between the nail, the plywood and the joist creates a squeak. Of course, adhesive helps keep the plywood and joist firmly connected. That should eliminate nearly all movement.

You'll also get squeaks if a nail just barely misses the joist and rides along next to it. As you walk over this nail, it rubs on the joist and squeaks. No amount of floor glue will eliminate this problem. Simply pull the offending nail.

By the time the building is ready for carpet, many trades will have walked over the subfloor. If any squeaks are going to develop, you'll hear them. On the day before the carpet goes in, walk the entire floor slowly with heavy steps. When you find a squeak, sink a few 16d nails in the vicinity. If that doesn't do the trick, start pulling nails.

The Pay

Sheathing and nailing the subfloor pays from 10 to 12 cents a foot. Sometimes the nails are provided, sometimes they aren't. The money is usually divided in half between sheathing and nailing. Obviously, the easiest money is in the nailing! A framer who takes eight hours to sheath 2000 feet makes around $12 an hour. If his buddy comes along with his gun and nails that same floor and three others that were done that day, and he does it in four hours, he makes about $100 an hour! Of course, not everyone can nail that fast. But even if it takes eight hours, it's obvious where the best money is being made.

Cutting Stairs

Stairs carry an unexplainable mystique. Just mention the act of stair cutting and you'll see seasoned carpenters scuttle off in embarrassed silence with their whimpering apprentices at heel. Journeymen roof stackers who can cut and stack even the most difficult roof will offer only blank stares when you ask for help in solving a stair framing problem.

It's a mystery to me why so many competent carpenters are so afraid of cutting stairs. Like cutting a wall or a roof, it's just a matter of having the right measurements and making accurate cuts. Although it's a job usually reserved for a journeyman carpenter or the boss himself, anyone who understands a few simple rules can master the basics of cutting and building a symmetrical set of stairs.

The function of a staircase is to provide access from one level to another. In theory this sounds straightforward, but too often a stairway has to conform to limited areas that won't allow for a simple solution. This is especially true in remodels, where you're stuck with existing openings. Many times you'll have to change directions, add landings or winders, or possibly incorporate curves or spirals.

Almost every set of stairs you build will present some design problems. That calls for a slower mind-set than most areas of rough framing. You'll probably have to work at the slower pace of a finish carpenter until you've worked out all of those problems.

We'll go through the basics of stair construction and even touch on some of the more complicated curved stairs. Since this is a book on rough carpentry, we won't go far into exposed open stringer stairways. Stringers for open stairways are usually laminated oak, built by finish carpenters. An excellent reference book on exposed stairway design and construction is *Designing Staircases* by Willibald Mannes (1982, Van Nostrand Reinhold Company, Inc.) which you can find in most technical or fine arts libraries. It explains laminating curved exposed stairways in detail. And I'd like to say thanks to the publishers for keeping the tricks to this dying art alive.

If your design calls for exposed wooden treads and risers (whether it's straight or curved), you'll need to consider that in your calculations. In this chapter, I'm assuming that your stairs will be carpeted so you'll just provide a rough plywood riser and tread. There wouldn't be any structural change if the stairs were wrapped with finish oak. You could use most of the exercises that follow for either kind of stairs.

But before we jump right in and ruin a stack of perfectly good lumber, there are a few ideas and terms we should cover. Here's the first, and most basic, concept: *We'll divide a finite amount of vertical space evenly, no matter what's going on between the bottom and the top.* It doesn't matter if you're cutting straight stairs, L-shaped stairs, or stairs that change direction 180 degrees at a landing: the measurement from the lower floor to the upper floor remains constant. So the goal is to divide the vertical space into comfortable, even increments. For layout purposes, any landing is treated as just another step. There can be any number of landings, in any number of shapes. That doesn't change the fact that all we're concerned about in the beginning is dividing the vertical space into equal amounts.

Stair Safety

The Consumer Product Safety Commission has determined that stairs are the most hazardous consumer product in the United States. An estimated 2 million accidents occur each year with almost 4,000 of these ending in a death. Your job as a carpenter is to make sure that any stairs you build are as safe as possible. But how?

People respond to a stairway using perceptual and physical reactions to the clues provided by the stair system. In other words, they approach a set of stairs with expectations. The most obvious are that the stairs will be uniform and that the handrail will support them if needed. After they climb a few steps, they let down their guard and expect their feet to take over. Any change in riser or tread dimension is liable to trip a person up, possibly sending them down the stairway. That's why each step should match its neighbors exactly in dimension and appearance.

If you build in blatant differences from one rise or tread to the next, and this causes a fatal accident, a court of law can hold you liable for this accident. It's not just a matter of losing your reputation. If you put up dangerous stairs, you stand to lose it all.

Always keep these factors in mind when designing a stairway:

1. The stairway should include at least three steps.

2. Sturdy guardrails should be provided on both sides of a stairway, especially on poorly lit stairways. They let people know they're approaching a stairway and emphasize any change in elevation as they descend the stairs.

3. Treads and risers should not vary more than ⅜ of an inch.

4. Install lighting that emphasizes the front edge of each tread.

5. Carpet or wood should be slip resistant, but not so slip resistant that you couldn't move your foot quickly to avoid a fall. Avoid using carpet with a busy design. Past legal battles have indicated that where a camouflage effect exists, negligence exists.

I don't mean to paralyze anyone with the fear of a team of cut-throat lawyers, but you are dealing with future liability. There are many phases to rough carpentry where you can walk away and be secure that any mistakes will be hidden when the house is finished. "Nobody, including the general contractor, will see that crazy bulge in the roof once the roof tile is up." But don't take this attitude with stairways.

Stair Terms

Figure 9-1 shows the parts of a typical staircase. The vertical part of the step is called a *rise* or *riser* rather than a step. The horizontal part (the part that you actually walk on) is known as a *run* or *tread*. These terms refer to the functions of the stairway, as well as the actual boards that you attach. I might refer to the rise of a step and then later refer to the attachment of the riser itself. The *total rise* refers to the height from the lower floor to the upper floor that the stairway covers. Since stairs always start and end with a rise, there will always be one more rise than run.

Code requirements vary, but always aim at using a rise of not much more than 7½ inches and a tread of not much less than 10 inches. If you have plenty of space, try to use a 7½-inch rise with a 12-inch tread.

The diagonal boards that support the stairs are known as *stringers,* although framers usually call them horses. Stringers should have at least 3½ inches of wood left after the treads and risers are cut into it. Usually starting with a 2 x 12 will give you enough material. Most stringers aren't exposed and are made with softwood such as pine or Douglas fir. When the outside stringer is visible and self-supporting, it's known as an open stringer. This is generally wrapped with finish hardwood such as oak. If your stringer is buried, it's called a closed stringer.

A *landing* is a flat platform that separates two or more flights of stairs. Landings are no more than small floors supported by short walls.

Winders are a collection of pie-shaped treads. These might occur throughout a whole circular stairway or just on the center landing separating one flight of stairs from another as shown in Figure 9-2. Winders have one end that's smaller than the stand-

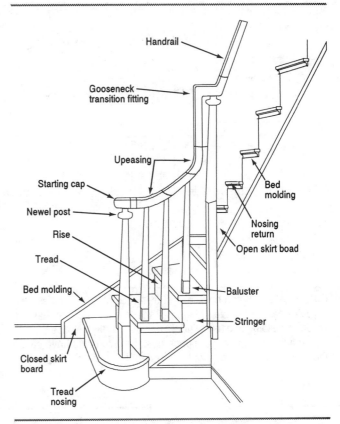

Figure 9-1   These are terms every stair cutter should know: rise, run (or tread), landing, skirt board, handrail, newel post and balustrade.

ard tread width. Generally the codes require that winder treads match the rest of the treads *along the line of travel* — approximately 12 to 16 inches in from the inside edge of the tread. Check your local codes. Some codes say the narrow end of the winder must be at least 6 inches wide.

Handrails run at the same pitch as the stairway. Outside exposed handrails are constructed with hardwood newel posts and balustrade uprights 4 inches apart. The inside handrail, sized so that a human hand can grasp it, should be securely mounted to the wall 30 inches above the nose of the tread. Leave a 1½- to 2-inch space between the handrail and any wall, including the brackets used to mount it. In dimly lit stairways the handrail should follow any changes in elevation (landings for example) to alert people with bad eyesight.

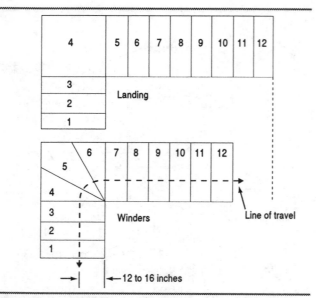

Figure 9-2   If you have a tight fit, winders can be a lifesaver. If you're having trouble fitting in all the treads, you may have to resort to building in a few winders on your landing. Notice that we saved two treads worth of space upstairs. We could have saved the same space downstairs by lowering the landing.

A *skirt board* is a piece of finish hardwood trim that's placed along the stairway walls and runs at the same angle as the stairs. The finish hardwood risers and treads finish up to this board.

Many codes require a head clearance of at least 78 inches. The *Uniform Building Code* requires 80 inches. Measure this head clearance by laying a straightedge or 8-foot level on the stairs under the header of the floor above as shown in Figure 9-3. Then hook your tape on the straightedge and measure up to the lowest point of the ceiling above, keeping the tape plumb. Some carpenters measure directly off the nose of the closest tread below. But this isn't what most building codes require. Always refer to your building code before you start to cut stairs. If your city or county has adopted the *Uniform Building Code*, refer to Section 3306.

While you've got the code open, check maximum rise, minimum tread width and depth, maximum difference in risers, minimum landing width

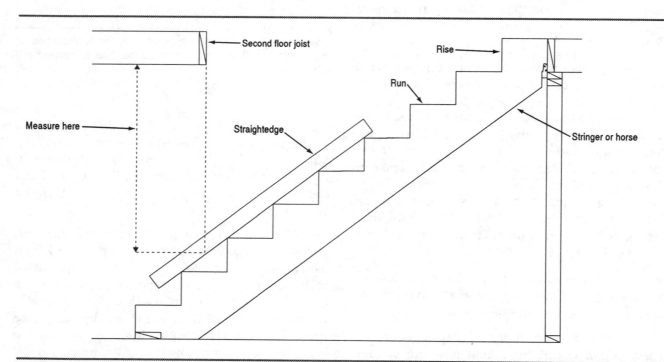

Figure 9-3   When measuring for correct headroom, be sure to use a straightedge set along the tops of the treads. A lot of framers just measure from the closest tread, but this isn't what the Uniform Building Code requires, or how an inspector would measure it. In tight spaces, this is very important. You don't want to tear apart your stairs later!

and depth, and minimum and maximum handrail and guardrail height. If you're building circular stairs, check the minimum tread sizes inside and out.

Stair Layout

There are many different types of stairways. In the following exercise, we'll work through the layout and construction of a stairway with a half-turn landing in the middle. But before we start, let me emphasize some general rules for building stairs. Always work carefully. Double-check your calculations before you do any cutting. If you're not confident, cut a horse out of plywood and check the fit before cutting a real horse out of 2-inch lumber. Make your mistakes on the plywood. If there's a problem, you haven't wasted a lot of time and lumber.

Practice your rise and run calculations on stairs already built. Compare your figures with dimensions of actual stairs. You can learn a lot from the mistakes of others.

Before cutting any lumber, make sure the design you're using will actually fit in the space available. Begin by referring to the plans to familiarize yourself with the design. Check the stairwell measurements (Figure 9-4). A few minutes spent planning the layout can save hours later.

Start by snapping out the location of the landing (if you have one) to the length, width and location shown on the plans. A rule of thumb is that the landing must be at least as deep as the stairs are wide. Most plans are drawn that way. If the stairs are 36 inches wide, then the landing must be at least that, and more if you have room. Where the stairs connect to the top floor is also considered a landing. Plan on at least 3 feet of width and depth between the top step and any opening or doorway, unless your code permits smaller landings.

Most importantly, make sure you have enough room for all of the treads. First check the treads on the lower stairs that approach the landing. If the plan shows seven treads before the landing, measure off

Figure 9-4   Before you can start building a set of stairs, the stairwell must be completely framed, plumbed and lined.

70 inches from the 36-inch landing line you snapped and see where it ends up. You may decide to go with larger runs later, but go with the minimum now, just to make sure it will fit. Are you encroaching into any hallways? If so, how can you fix this problem? Do you have room to transfer one rise to the upper stringers? How about putting a few winders on the landing?

You might have to incorporate quarter-turn winder stairs to save room. If you can use up a few risers in the space that would normally be a flat landing, you have that many fewer treads to worry about. See Figure 9-2. Winders will compact a design so that the whole stairway itself takes up much less room. Flat landings are much safer than winders, but if space is tight, you might be left with no other choice but to go with a few winders.

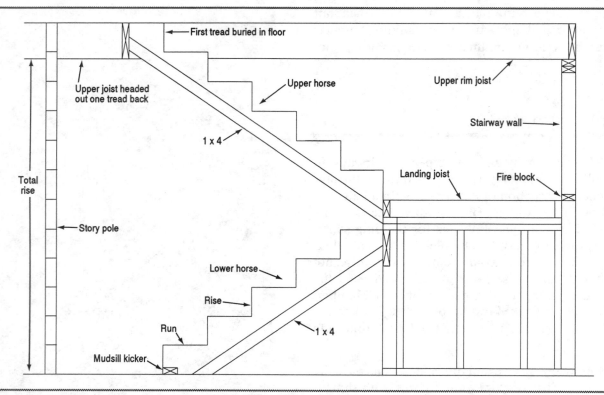

Figure 9-5   To double-check your work as you go along, make a story pole. Simply cut a piece of wood the same height as the overall rise, and then lay it out along its length with your calculated rise. Notice how the layout markings line up with each rise on the horses. The 1 x 4 on the lower sides of the horses keeps the horses from being attached tight to the handrail walls, yet still gives you a solid place to attach the handrail to the horse. Leaving this space is an act of courtesy toward the drywallers, so they don't have to notch their drywall around every rise and run. Make sure to compensate for this additional 1½ inches in the rough opening when laying out your stairs.

Once you're satisfied with the lower stringers, check the upper stringers that lead from the landing to the second floor. This is where most problems occur. Many times the upper floor joists were framed with little or no thought to the stairway. If the plans show seven treads between the landing and the upper floor, you will need *at least* 70 inches here and maybe more depending on how you plan to attach the tops of the horses to the upper floor joists. I like to design my stairs with an extra tread on top, which gives the best connection between the staircase and the floor framing. More on that later.

If the opening from landing to upper floor joist is too small, you've got a problem. At this point you have only two choices. The first is to reframe the upper joists. The second is to change the design of the stairs and incorporate a few winders on the landing. You usually can't do this in new construction, but with remodels, "you gotta do what you gotta do."

Once you've established that the stair layout will fit the available space, you can calculate the exact rise and run. It could vary from the dimensions the designer planned.

■ Riser Height

To calculate the height of the risers, begin by measuring the floor-to-floor height. That's the *total rise* of the stairway. See Figure 9-5. Let's suppose the wall height is 9'1", with 2 x 10 floor joists and ¾-inch sheathing. The total vertical rise would be

119.25 inches (109 + 9.5 + 0.75 = 119.25). Each individual rise should be about 7.5 inches, so we'll divide the total rise by 7.5. If you divide 119.25 by 7.5, you end up with 15.9 rises. Now obviously you can't have less than a whole rise. So you'll have to decide whether to go with 15 or 16 rises. How do you decide which?

If you divide the total height (119.25 inches) by 15, the rise is 7.95 inches. Divide by 16 and the rise is 7.45 inches. Sixteen rises at 7.45 inches is a much more comfortable step, so we'll use that for our sample stairs. But when you're building stairs in a limited space, a small change like that can make a big difference, as we'll soon see!

Check the design of the stairs shown in the plans. Does it show 15 or 16 rises? Remember, the landing face is a rise. More importantly (for stairs designed in tight areas) does it show 14 or 15 runs? If you only have room for 14 runs, you may have to go with the taller rise. This is where you will fine-tune the layout to make sure that what you have *on site* matches what the designer planned.

There are a few formulas which express the relationships of the rise and tread dimensions you should keep in mind. The first was introduced in 1672 by Francois Blondel of the Royal Academy of Architecture in Paris:

$$2 \times R + T = 24 \text{ to } 25 \text{ inches}$$

Other formulas used in the past include these:

$$R + T = 17 \text{ to } 17.5 \text{ inches}$$

$$R \times T = 70 \text{ to } 75 \text{ inches}$$

I'm not a big fan of formulas. These formulas are way too ambiguous for me. If you use a rise of 7.45 and a tread of 11.75, in the first formula you miss by over 1½ inches (7.45 + 7.45 + 11.75 = 26.65). Plug those same numbers into the last formula and you miss by over 12 inches! But 7.45 by 11.75 is a very typical and safe combination!

Modern codes have eliminated these formulas from their texts. They're a nice bit of history for every stair builder to know, but don't rely on them. They exclude too many otherwise completely safe designs.

At this point, some framers like to make what is known as a *story pole*. It's easy to make, and helps to double-check your calculations. Simply cut a board to the same length as your total rise. Then lay out the height of each individual rise on the story pole as shown in Figure 9-5.

In our example, we would mark off exactly 16 rises of 7.45 inches. If it doesn't come out right, something's wrong. If it works out, set the story pole aside and continue. You'll use it again later to do some double-checking, so keep it handy.

■ Tread Depth

Now we're ready to complete the last step in fine-tuning the layout of the stairs. Remember, there will always be one more rise than run. If there are eight rises on the lower stringer, there will be seven runs. If you're building stairs in a limited area, this is the most important part of laying them out.

We know that the typical 10-inch run works for our stairs, but let's assume that the plan calls for 11-inch treads. Seven runs at 11 inches each equals 77 inches. Now recheck the proposed stairwell to make sure that 77 inches from the face of the landing will fit without running out into a hall or encroaching into a room. Check the plans to see if your stairs will end where the plans show them ending.

In the photographs accompanying this chapter, the stairs had to end exactly where a radius handrail ended. The architect was adamant about keeping the radius exactly as the plans showed. When calculated with 11-inch treads, the stairs ended 77 inches out from the face of the landing — which was about 1¾ inches too far. By reducing the treads from 11 to 10¾ inches, I reduced the length to 75¼ inches, saving the exact 1¾ inches I needed. Then I cut the risers on an angle, squeezing out another inch in tread depth. So the treads were

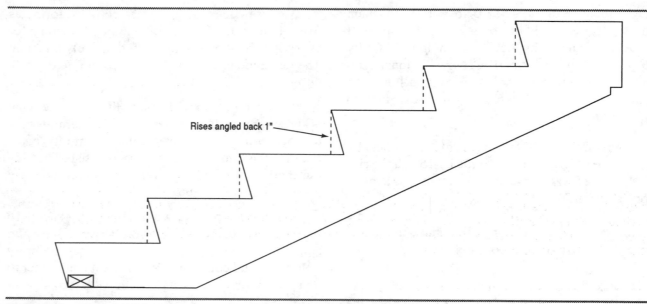

Rises angled back 1"

Figure 9-6   A good trick for squeezing out an extra inch of tread space is to cut each rise at an angle. While you're laying out your horse, measure back into the horse an inch on each tread. Then draw a new line from the nose of each tread back to this inch mark to create the angle.

11¾ inches in the end, an adequate tread. Figure 9-6 shows how cutting the risers at an angle increases the depth of the tread.

■ Heading Out the Upper Floor

We're nearly ready to cut some stairs now. We know that the treads on the lower stringer worked at 10¾ inches. This totaled 75¼ inches for seven treads. Let's see if this works on the upper stringer. Verify the *actual* distance from the proposed landing face to the upper floor joist that the horse will attach to. I was around when this particular floor was headed out, so I had calculated exactly where I wanted it placed during its construction. This isn't often the case, though.

When I lay out the floor head-out (that is if I get there before it has been done), I usually add one more tread to the count. In other words, instead of allowing 75¼ inches from the face of the proposed landing *to* the head-out, I'd add one more tread, for a total of 86 inches *to* the head-out. Measure by leveling up from the face of the proposed landing, making a mark on the stairway rim joist. Then measure away from the landing 86 inches and have the floor headed out here.

Framing for the extra tread on top makes for a more solid connection between the horse and the floor joist. See Figure 9-7a. If you use the floor head-out as the last rise, you're eliminating the possibility of using a hanger on each horse. This is because the top of the horse is hitting the floor joist 7.45 inches down from the top of the floor sheathing. If it's a 2 x 10 joist, that only leaves 2 inches worth of joist to connect your horse to! Some people solve this dilemma by hanging the horse off the joist with a steel strap or a strip of plywood as shown in Figure 9-7b. But that's only a glorified band-aid in my book. If there's a wall supporting the stairs under this upper connection as shown in 9-7c, then attaching a hanger is no problem.

If you can have the upper floor headed out one tread farther back as in Figure 9-7a, by all means do it. You'll be building a much stronger set of stairs. The floor sheathing upstairs covers the extra tread like it was never there.

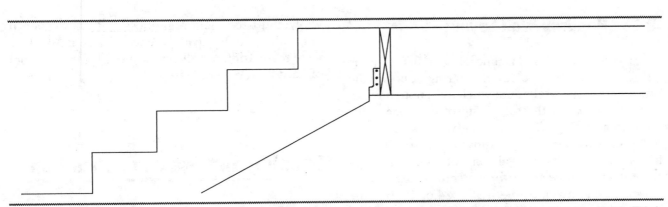

Figure 9-7a   There are three ways to attach a horse to an upstairs joist head-out. This floor was headed out one tread back. This is the best connection you can make, as it leaves plenty of room for a hanger. When calculating the floor head-out, just add an extra tread (or even half a tread) to your measurement, and head it out there. Notice how the last tread is level with the top of the floor joist. When the floor is sheathed, the tread disappears from view.

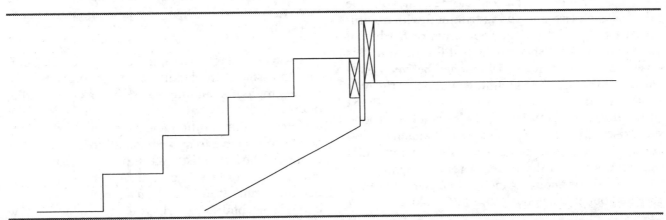

Figure 9-7b   The alternative is to frame your head-out exactly where the last rise lands, and then connect the horse with a length of plywood or a steel strap. If you choose to go with this method (or are forced to), don't forget to add a 2 x 6 block between each horse. Nail the block into the horse and then send a few nails into the joist head-out.

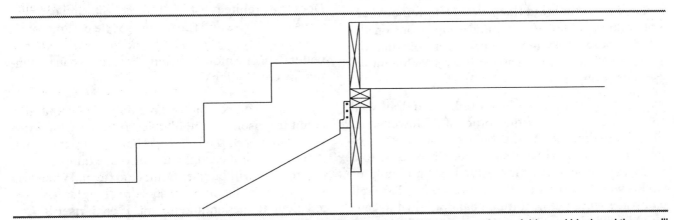

Figure 9-7c   If you have a supporting wall directly under your joist head-out, you'll be in better shape. Add a pad block and then you'll have some meat to hold your hanger securely. This illustrates the top connection of a lower stringer and landing connection as well.

■ The Handrail Wall Layout

If you have to build a handrail wall between the stringers or on the side of either stringer, now's the time to lay it out. Make sure you leave enough width between walls for the full width of the treads plus the wall coverings. The width between walls has to be at least 37.5 inches to allow room for 36-inch treads and ⅝-inch drywall on both sides. *Always remember to leave space for wall coverings.* Here's a trick I use: Attach a length of 1 x 4 or 1 x 6 to the lower edge of each horse before you nail it in place. (You can see the 1 x 4s in Figure 9-5.) That guarantees the drywallers enough room to slide wallboard between the edge horse and the wall.

In better-quality custom homes, the outside edge of each rise and run is trimmed with a hardwood 1 x 12 skirt board. The carpet or hardwood riser and tread finishes up against the skirt. In this case, you have to leave 1½ inches between each outside horse and the handrail walls so that the wall coverings and the skirt board can slide down next to the horse, without having to be notched around each rise and run. In this case you'll attach a 2 x 4 to each horse instead of the 1 x 4. Keep in mind that your stairs would then need a rough opening between walls of about 39 inches to ensure that they finished out at 36 inches (36 + 1½ of skirt + 1¼ of drywall).

When you lay out the handrail wall, you just mark out the location of the bottom plate on the slab. Then you're sure you aren't encroaching on the wall space when you build the stairs. Determine where you want the wall located and snap it out on the floor. These lines are essential for locating the position of the lower stringer floor attachment, so get them down now.

In our example, we had a radius handrail wall located on the bottom right corner of the lower stairs drawn on the plans. The builder had dreams of putting an exposed handrail in its place so he told us to leave the wall off. I pulled out a couple of radius wall lines on the floor for the radius handrail as I was laying out the middle handrail, just in case he changed his mind later. As you'll see farther on, we built each rise and run around these marks, as

we would need them regardless if there was a rail or a wall there. In the end the wall was built, and we were ready for it. Chalk up a victory for this "school of hard knocks" graduate!

Landing and Stair Construction

Let's pause to take stock of what we've done so far. We've found the optimum number of risers and located the length and width of the landing. We checked to make sure that the rise allowed for the appropriate number of runs to fit in the space we have to work with. Then we snapped out the handrail walls.

The next step is finding the height of the landing. In essence, the landing is just another rise. In our example the landing was drawn as the middle, or eighth, rise. My calculations matched that of the architects. With a little luck yours will too! If they don't, you'll have to make the additional runs fit or change to the larger rise. At this point you have three options: Is there room on the first set of stairs for one more run? If not, can you squeeze another run in on the second set? Or do you have to build a few winders on the landing?

But before you make any drastic design changes, try to figure out why there's a difference. Does the architect list the wall heights and calculations for his rise and run design? Is he calling for an unusually high or low rise? One less rise is always better than one more rise, but why aren't you matching his calculations?

When you're comfortable enough to continue, count the risers on the plans to find out which rise is the landing. If the landing is halfway up the stairs, it will be number eight (as in our example). The upper floor will be riser number sixteen. When you know that the landing occurs at riser eight, it's easy to calculate the landing height. If each riser is 7.45 inches, just multiply 7.45 times 8: 7.45 x 8 = 59.6, or about 59⅝ inches.

I strongly recommend writing out all your calculations on a wall, even though you're using a calculator. If trouble arises later, you'll be glad to have the numbers to refer back to.

Lay out the landing on the lower floor so you're sure the stairway fits in the space available and according to the plans. When that's done, you can feel confident enough to start cutting some sticks.

The landing is actually a small floor supported with a face wall and rims against the stairwell walls. For a small landing, as in Figure 9-8, 2 x 6 joists are more than adequate. To find the size of the studs for the face wall, you've got to work backwards. The floor sheathing is ¾ inch thick, the joists are 5½ inches, and three plates for the wall total 4¾ inches; this adds up to 11 inches. When you subtract 11 inches from the total landing height, you get 48½ inches. When building the face wall, use 48½-inch studs, and then build the landing floor on top of it.

In some parts of the country, 2 x 4s are exactly 1½ inches thick. In others (California, for one) 2 x 4s are milled closer to 1⁹⁄₁₆ inches. So three plates would equal closer to 4¾ than 4½ inches (in case you were wondering how I got the 4¾ inches for the three plates.)

On the front face of the landing where the lower horses attach, install 2 x 10 blocks flat to the outer face of the wall as shown in Figure 9-8. These provide a nailing surface for the hangers. I like to delay sheathing the landing until I install all the treads and risers. At that point I'll have the plywood, glue and nail gun out, and can do it all at once.

Once the landing is built, get your story pole and double-check the landing before you continue. Find the eighth mark up the pole and make a mark ¾ inch below it. This mark should line up with the top of the unsheathed landing. If it doesn't, something is terribly wrong! The story pole, if you've stepped off the proper rise markings, will never lie.

Many carpenters who have only been cutting stairs for a short while don't like to waste time with a story pole. They let the old men mess around with these rigs. They would rather get the stairway farther along before they find their problems so that they can waste much more time and material.

Figure 9-8   Here's the middle landing completely framed and joisted. A landing is actually only a small floor supported by a face wall. Notice the 2 x 4 fire blocks in the back stairwell wall. There are still a few missing in the left corner.

■ Laying Out the Horses

When everything is checked and double-checked, it's time to cut the horses. The horses are most commonly made from 2 x 12 Douglas fir set 16 inches on center. But always check what's shown on the plans. Many people will use three horses for a 3-foot wide stairway. I use four. I'm often accused of overkill. Then again, I'm never accused of springy stairs.

Select lumber for the horses that's a couple of feet longer than you'll actually need. The ends of long pieces of lumber are usually split. Plan to cut off each end so the horses don't break along these

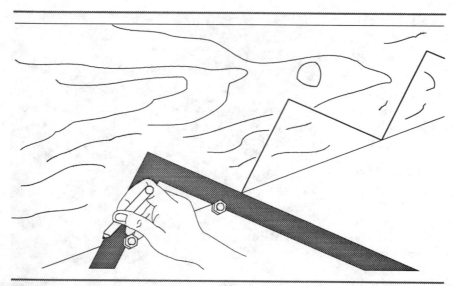

Figure 9-9   Stair gauges on the framing square help hold the rise and run dimensions exactly on the marks. Notice that the framer is tracing around the outside of the square.

splits. With the rise and run marked out, you will want 3½ to 4 inches of wood left on the face of the stringer to ensure adequate strength.

You can mark out the rise and run on the horse with a template cut out of a piece of plywood, or with a framing square. To make a template, simply cut a triangle-shaped piece of plywood with one side the length of the rise and the other side the length of the run. Then hold the hypotenuse (or pitched) side of the triangle along the crowned edge of the horse stock while making each rise and run mark. Trace along the rise and run and then slide the template up the horse until the point of the run on the template meets the last rise mark laid on the board. A template is very accurate, as you'll always trace the exact same pattern.

The most common way to make your marks is to use a framing square. Lay the square on the horse stock with the heel on the board and the tongues pointing off the board. On the outside edge of each tongue are inch markings; from 1 to 16 along one tongue and 1 to 24 along the other. One tongue would represent the rise and the other would represent the run. The square has inch markings on the inside and outside of each tongue but always use the outside markings on both tongues. In other words, don't use the outside marking on one tongue for your rise and then the inside marking on the other tongue for your run. It won't work! Stay with the outside markings on both and you'll do just fine.

The main problem with using a square is that you might get careless when you're marking each rise and run. That makes consistency unlikely. This problem is easily solved by using a set of stair gauges (brass fittings that tighten down on your square). Hold your square with the rise and run on the outside edge of each tongue lined up along the crown side of your board, and then tighten the gauges snug against the board, along the inside edge of your square. After you mark the first rise and run, you simply slide the square up the crown edge until the next run matches with the last rise marking, then mark the next rise and run. See Figure 9-9. Move up the crown edge until you've marked all the rises and runs.

The landing itself is rise number eight, so you'll have to mark out seven rise and run combinations on each horse. Be sure to mark them on the crown side of the board, so all the crowns face up when the stairs are up. Once you've marked them all, there are two points on the horse that need adjusting — the first rise and the top run.

If you marked out everything correctly, the first rise should measure 7.45 inches from the finished floor level to the toe of the first tread. But what happens when we add the ¾-inch rough tread to the first run? The rise grows to nearly 8¼ inches. And what if the finish carpenter adds a ¾-inch piece of oak? You would have a 9-inch rise! To compensate, trim ¾ inch or 1½ inch (the thickness of the total tread material, rough as well as finish) off the edge of the horse that rests on the floor.

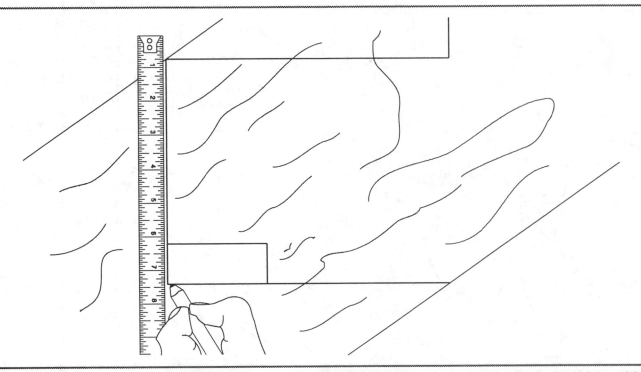

Figure 9-10   This is the first rise and tread. The first rise will always need to be adjusted. "Drop" the first rise by the thickness of the tread. Otherwise the first rise would be ¾ inch (the thickness of the tread) taller than all the rest. *Burn* ¾ inch by holding your tape ¾ inch above the mark for the first run. Then you can mark out the rise exactly on your tape, without having to do any calculations. In this case the rise was 7.45 inches. You're pretending to hook your tape on the upper edge of the first *finished* run, then marking the calculated rise to find the line to cut for the horse to sit on the floor. Always adjust the floor cut rather than the first tread cut. Otherwise you just transfer the problem one rise up. Notice the 2 x 4 notch for the mudsill kicker.

Don't mess with doing the math to find your height, just burn ¾ inch on your tape above the first rise as shown in Figure 9-10.

Never cut the excess from the first run; this only transfers the problem one rise up. Always cut the line that actually rests along the floor. While you're down there, make a 1½- by 3½-inch flat notch in the lower front of the first rise. This notch accepts the 2 x 4 mudsill kicker that attaches the front of the stairs to the concrete. More about that later.

Remember, I said there are two adjustments. You make the second one to the top run. If you marked out the horse properly, there's a 10¾-inch top run. When you nail on the last riser, the width of that run grows to 11½ inches. To compensate,

take ¾ inch off the back side (the face opposite the final rise) that attaches the horse to the landing. See Figure 9-11.

Again, don't mess with any math, just burn the ¾ inch with your tape. Then make another 2-inch notch at the bottom of this line to accept the hanger. If you're adding a piece of finish oak to each rise, now is the time to take this into consideration. You'll need to trim another ¾ inch to compensate, so that every rise is the same in the end.

If you wanted to cut your horses with an angled riser like I did on this particular set of stairs, now's the time to make this adjustment. See Figure 9-6. Start the adjustment at the toe edge, the point formed where the rise and tread meet, and angle back into the stair to where the rise lands on the

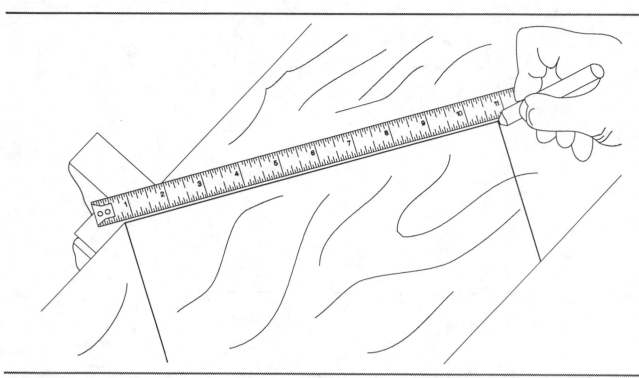

Figure 9-11   Here we're adjusting the top tread so it ends up equal to all the others when the stairs are complete. The depth of the top tread has to be cut shorter than all the other treads in the stairway. Otherwise, when the top riser is installed, the tread will end up deeper than all the rest. We're again burning ¾ inch as we mark the tread size, which is 10¾ inches. The line that's being marked out is the vertical face parallel to the last rise. This is the edge of the horse that will be nailed to the face of the landing. Always adjust this edge rather than the edge of the last rise. Otherwise you just transfer the problem one tread down.

lower tread. Simply mark out 1 inch on each lower tread and then make a new riser marking from the top of each rise/tread toe junction down to this inch mark. This will angle the rise and make the tread 11¾ instead of 10¾.

Keep in mind that this will affect the second adjustment on the horse that we spoke of previously. If you made your last tread 10¾, extend the mark to 11¾. Then, *don't forget* to subtract ¾ inch from this tread along the backside of the horse that attaches to the landing so you end up with a tread that matches all the rest. Whew — are you still with me?

Don't let your mind slip into idle when you're cutting stairs. Think through every step. Double-check your procedures. Measure twice and cut once. Work carefully and you'll build the confidence and reliability every master carpenter needs.

■ Cutting the Pattern Horse

When you've measured and double-checked the adjustments, begin cutting the first horse. When it's cut, lay it in place and check the fit. Measure from the top of the unsheathed landing down 7.45 inches on each end, and snap a horizontal chalk line along the face of the landing at this point. See Figure 9-12. This is where you'll want to nail the top of each horse. Tack the first horse up at this point.

Now, does it fit plumb against the landing? If it's out just a little, you may want to adjust the top vertical cut. If it's out a lot (as much as ½ inch), you probably did something wrong. Did you measure down from the *unsheathed* landing 7.45 inches? Remember the top tread is also unsheathed at this point. Are all the rises and runs cut the same, except

Figure 9-12   A snapped line on the landing gives you a reference for nailing the stringers. When toenailing material like this, always start your material above the line and let the nail and your missed swings drive the wood down to the line.

Figure 9-13   Once the two outside lower stringers are set, shoot a piece of mudsill in place. This "kicker" will secure the front of the stairs.

for the adjustments to the first rise and last run? Is the landing built plumb? Check the bottom of the horse. Is it level with the slab?

Backtrack until you find the problem, then work out the solution. With the first horse in place, the rises and runs should be plumb and level, unless you cut the risers at an angle. Also check to make sure you have at least 6′8″ head clearance after the drywall is installed.

Be sure the first horse is correct before you cut any more. When it's right, use the first horse as a pattern to mark all the other horses. Also mark out an extra horse for the upper flight of stairs, but don't cut it! The upper horse will have different adjustments built into it.

For a 3-foot wide tread, I like to use four horses at 12 inches on center. Of course you could get by with three horses at 18 inches on center, but four horses make a stronger stair. On a tract, the plans will probably call for horses set at 16 inches on center. It's up to the builder. On a custom, the framing contractor will make the call.

◼ Hanging the Horses

After you've cut all the lower horses, begin hanging them. If the stairs won't have a finish skirt board applied by the finish carpenters, nail a 1 x 4 on the outside of the two side horses to create a gap of ¾ inch between the wall and the stairway. See Figure 9-5. This will leave room for the drywallers to slide their material down the sides of the stairs. If a skirt board is to be applied later, use a 2 x 4 instead. Nail up all the horses so their top treads are on the line you snapped earlier (7.45 inches down from the top of the unsheathed landing). See Figure 9-12.

Now slide a piece of 2 x 4 mudsill that's the same width as the treads (not counting the 1 x 4) into the slot you cut. Shoot it down into the slab or deck, making sure it's ¾ inch away from the handrail layout markings on the floor. See Figure 9-13. Nail the horses to the mudsill kicker, adjusting them with shims if needed, as shown in Figure 9-14.

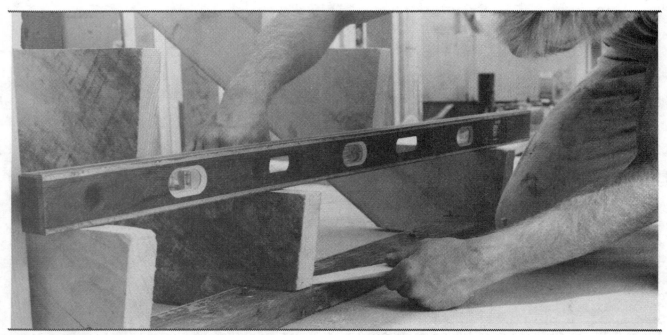

Figure 9-14   A straightedge and some wedges will help to keep everything in line.

Figure 9-15   The handrail for this set of stairs was designed at a slight radius. The contractor hadn't decided whether to go with a wall or an open rail when I was building the stairs. I decided to run everything plumb with the wall line I had drawn on the floor. I always install the risers first and snug them up against the tread if they are cut a bit short.

Once the horses are up, measure and cut plywood for the risers and treads. The grain of the plywood should run perpendicular to the horses. Put the risers on first, holding them up to the top of the rise if they're a little short. This makes for a cleaner-looking job because the treads will cover the gap at the bottom of the rise. Figure 9-15 shows the risers nailed to the horses. On this job, we had a radius handrail to be framed later. That's why the lower rises run wide to the right — they're following the radius drawn on the floor.

Always run the tread over the top of the riser. Apply glue generously, then nail the treads and risers with 8d nails, as shown in Figures 9-16 and 9-17. Put at least three nails through each tread and rise into the horse. Try working your way up the stairway doing a few rises and then a few runs so that you don't risk a fall. See Figure 9-18. Once you've completed sheathing the lower rises and runs, sheath the landing and nail it off as you would nail a floor.

Where the stairs run along a wall or handrail wall, you'll need a fire block running along the angle of the stair against the 1 x 4 mounted to the

Figure 9-16   I use a quart-size glue gun to spread a thick bead of glue on each rise and run.

Figure 9-17   With a nail gun, or by hand, use 8d nails to secure the risers and treads. Put about three per rise and three per run into each horse.

horse. See Figure 9-19. Add another fire block in the stairwell wall around the top of the landing, as shown in Figure 9-5. These keep a fire that starts under the stairway from shooting up inside the adjacent wall. Inspectors love to call you on these if you leave out even one. If there won't be a skirt board installed, add 2 x 10 flat pad blocks where the point of the tread ends against the wall. That helps protect the wall so people climbing the stairs don't accidently kick through the drywall.

That finishes the lower portion of the stairs (the part below the landing). The upper portion is nearly the same. If you allowed for one extra tread on top as I recommended earlier, add it to your pattern now. See Figure 9-7a. Then you'd have eight risers and eight treads, with the last tread covered by the upper floor sheathing. Make a notch in the lower portion of the face that attaches to the joist so you can get a hanger in there later as shown in Figure 9-7a.

The lower attachment on the upper stringer is also different. Instead of having a flat surface parallel to the first run for the horse to rest on, carry the

Figure 9-18   As you work your way up the stairs, do a few risers then a few runs. That way you have something to walk on as you go up.

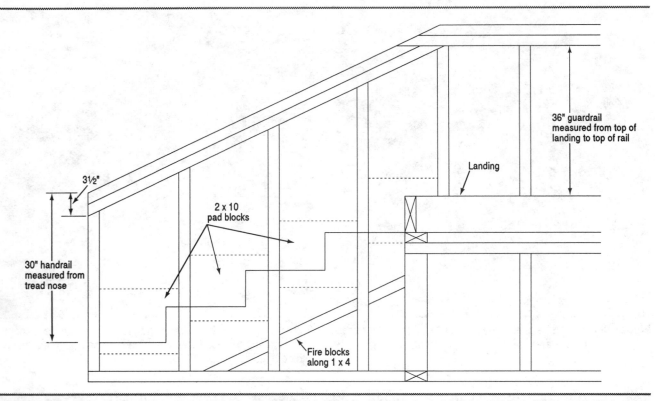

Figure 9-19 Most codes call for handrails to be 30 inches tall. When measuring the height of your handrail, measure off the nose of the tread to the top of the finished handrail. Guardrails around landings or upstairs hallways should usually be 36 inches (but check your local codes). Notice how the handrail top plates measure 1¾ inch when set on an angle, not 1½ inch. Make sure to get your 2 x 4 fire and 2 x 10 pad blocks in.

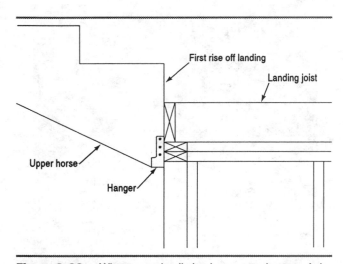

Figure 9-20 When you detail the lower attachment of the upper horse, carry the rise mark all the way down, because this face attaches to the face of the landing. Put a flat cut on the bottom to accept a hanger.

first rise marking all the way down until you run out of wood to mark on. See Figure 9-20. Then put a 1½-inch flat cut on the bottom to allow for a hanger to support the horse. This extended riser line will fit flush against the face of the landing.

As before, always cut a pattern horse and then adjust it, if necessary, before cutting any more horses. This stopping to check the pattern has saved miles of wood even in my relatively short career.

When attaching the horses, always use a straightedge or level to keep the horses even. See Figure 9-21. Attach the two outside horses and then attach the inside horses when they fit snug up against the level. While you're there, also check the face of the risers.

Figure 9-21   When attaching the upper stringers, use a straightedge to keep everything in line. These stringers had to be cut back around the closet wall that's supporting them to keep them from protruding down into the closet ceiling. Notice the large notch directly beneath the first tread down from the upper floor. The location for the vertical face of the notch was found by measuring off the face of the landing *to* the face of the closet wall. After cutting them, I glued and nailed plywood gussets to strengthen what was left of the stringer. Notice how the upper floor head-out was moved back one tread to strengthen the attachment. Also note the ¾-inch gap between the right stringer and stairwell wall. This gap will be filled with drywall.

Building Winders

Winding stairs are permitted by most codes if you observe certain minimum dimensions. Winders save space and, if built properly, can be almost as safe and as comfortable as a conventional stairway. See Figures 9-2 and 9-22. Figure 9-2 shows a quarter turn winder and Figure 9-22 shows one way to work out a half turn winder. The idea here is to split up the landing into an even number of pie shapes, depending on the size of the landing and the number of treads you need to fit on the landing.

Winders shorten the run on the upper or lower flight of stairs, so you can compact the whole design. In tight quarters they're a lifesaver for the designer. But they can be more dangerous for the user because the treads vary in size. Some codes allow the treads to come to a point as they turn a radius, while others stipulate at least 6 inches worth of tread at the narrow end.

You may also need at least 10 inches along the "line of travel" on the stairway, the line a climber follows when going up or down the flight of stairs. The line of travel is typically around 12 to 16 inches in from the narrow end of the tread. Check your local codes — and keep these dimensions in mind when figuring out your winder sizes.

Calculating the size of each tread can be tricky. Since the winder size depends on the space you have, start by just dividing up the landing into equal pie shapes and see what works. Lay out the landing size on a spare piece of plywood, or on the floor somewhere, and start experimenting. Try cutting a few 2 x 4s and use them as guides. Move them

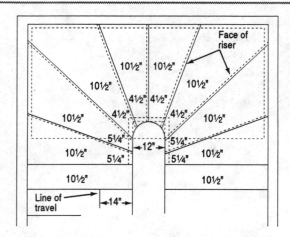

Figure 9-22   This is a half turn winder. Notice that when laying out the winding treads the *line of travel* was kept at the same depth of tread as that on the common treads.

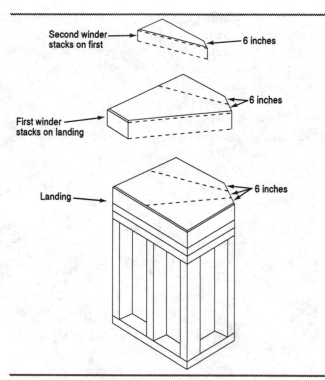

Figure 9-23   Each winder is actually a box that's stacked on top of the winder below it. The material used to frame the boxes is ripped down to the height of the rise minus the tread material.

around until you have even treads and the width at the line of travel is correct. Or, if you're lucky, your plans will have detailed drawings with all the numbers you need to work with.

Winders are actually small boxes that are built and placed directly on top of the landing, once it's constructed. See Figure 9-23. The first tread of the winding section of stairs is that portion of landing that wasn't covered by the first box. The boxes are built and stacked on top of each other in the shape that you design. Each box or tread will cover up most of the box beneath it, leaving only the lower tread exposed beneath it. The material you use to build the boxes must be ripped down to the height of the rise *minus* the height of the tread material. Each winder design and size will vary with the special circumstances of the job.

Building the Handrail Walls

I like to build my handrail walls *after* all the stringers are up. But if I lay out the handrail walls before I put up the stringers, I'm sure there won't be a problem with the stringers encroaching on the space that the wall needs.

Most handrail walls will be built on a rake, following the exact pitch of the stairs. If you stick frame the walls you won't have to do any figuring for rake long and short points. Most codes call for handrails to be built at a height of 30 inches, from the nose of the tread to the top of the handrail. See Figure 9-19. A railing around a landing or an upstairs hallway is considered a guardrail. Most codes require a guardrail to be 36 inches tall.

If the handrail wall is detailed with a finish piece of hardwood or a rounded wood or steel handrail that attaches to the top of the wall, be sure to consider this in your measurement. Subtract anything that attaches to the top of the rough-framed wall from the 30 inch measurement.

To begin the wall, attach a bottom plate to the floor. Check your plans to see where the handrail ends. Usually it's in line with the first rise of the stairway. End the bottom plate at this point. Next, lay out the bottom plate for studs at 16 inches on center.

Now you can figure out the height of the first stud. Measure the distance from the bottom plate to the top of the first rise. Then add about 36 inches. This will give you a stud taller than you'll need, but you can cut it down later. After you cut the first stud, nail it down to the bottom plate and then fetch your level. Plumb up the stud and then toenail it into the 1 x 4 that's attached to the horse. Then repeat this process for the last stud. Use the same method to fill in all the studs in between, making sure that the top of each stud is about 36 inches above the nose of the tread it stands nearest to.

When all the studs are plumb and nailed to the 1 x 4 on the stringer, cut them to the proper height. You'll have a line of studs nailed to the bottom plate and horse with their tops ending roughly 36 inches above and parallel to the line of tread noses.

Now you need to mark out the first and last stud appropriately and snap a line from one to the other. That gives you the angle and height to cut all the middle studs. You'll have two top plates on top of these studs so subtract their height from the 30 inches. Look at the lower end of the handrail in Figure 9-19. Each plate set along the pitch of a typical stairway is around 1¾ inches measured plumb along its vertical height. So: 30 - 3.5 = 26.5. If you have anything else going on top, make sure to subtract it now.

Measure from the nose of the tread up the edge of the first stud 26½ inches (or your calculated height) and make a mark. Then do the same on the last stud. Then snap a line from the first to the last stud which will make an angled line along the edge of all the middle studs. Now take your square and draw a line from the long point of the line we just chalked, along the face of each stud. Take your speed square and find the angle that the line has laid along the edge of your studs. This is the angle that you will want to set your saw when making the top cuts of the studs. Go down the line and cut the top of each stud to this angle.

Chase down a couple of 2-bys for the top plate material. Lay down the plate on the cuts you just made on the studs, and tack it to the top stud. Now mark the two ends by tracing up the angles on the two end studs. Cut the top plate along these angles,

Figure 9-24   The builder on this project wasn't sure if he was going to go with an exposed open handrail or a wall here, so I decided to run everything along the radius drawn on the floor just in case. Each tread and rise followed the radius on the floor.

and then use it as a pattern for the other top plate, after you've checked it for fit. Put two 16d nails through the first top plate into the tops of each stud. Then nail the second top plate securely to the first.

If you have a radius handrail, build it the same way. I prefer to stick frame radius handrails, because there's a lower and middle attachment to hold the studs in place. Then it's just a matter of creating a smooth cut line along the middle studs from the first stud to the last.

If you have a radius handrail that ends at the bottom of the lower stringers, you might need to have the first tread end on a radius also. Figure 9-24 shows the first tread ready to be cut. Lay out the radius on the floor and on the first tread, and then

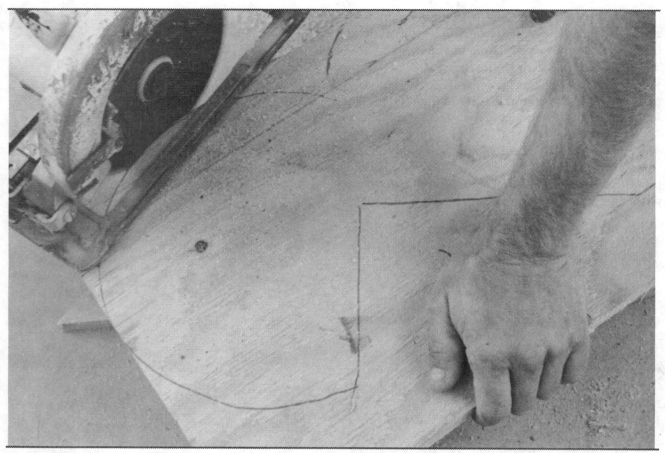

Figure 9-25   The first tread had to have a circular end regardless of the design of the handrail. Since the tread depth was 1¾ inches, the radius of the circle drawn was 5⅞.

cut the tread at the appropriate radius as in Figure 9-25. Check your cut using a square laid on the floor as in Figure 9-26. Then build a small radius wall under the tread to support it.

The handrail wall in Figure 9-27 was built after the contractor decided to abandon his idea of an open handrail. It was stick framed just as I described, then drywalled.

Circular Stairs

Circular stairs can be built with closed or open stringers. A closed stringer is built with small walls that create each step. The space below the stairs could be closed off and used as a storage closet. Open stringer stairs are self-supporting and are completely open underneath them. Open stringers are often laminated from ¼-inch pieces of oak and finished out at 1½ to 2½ inches. The entire stairway could be built by finish carpenters in a custom shop then reassembled on the job site, or built on the job site to begin with. Let's walk through the construction of a circular stairway with a closed set of stringers.

The first step to building a closed stringer set of stairs is to mark out the radius or circle that the stairs will follow, directly onto the floor. The plans will specify the inside as well as the outside dimensions to follow. The inside curve of the stairs will be a tighter radius than the outside, but both will usually be pulled from the same point.

Figure 9-26   Set your square on the floor to check the cut on the tread with the marks on the floor.

Figure 9-27   In the end, a handrail was built and then drywalled. At least we were ready for it!

To detail the curve on the floor, start by driving a nail into the floor on the center mark. This is the exact distance away from the outside radius that the plans call for. The center mark is the spot where lines measured in the same distance from both perpendicular square stairwell walls intersect. Look at Figure 9-28. Hook the end of your tape on the nail. Hold the pencil alongside the appropriate number on your tape and simply drag it along the floor while you mark out the area on the floor where the stairs will be constructed. Your tape will rotate on the nail and as you walk backwards you'll mark out a perfect circle. After you mark the outside radius, check the dimension for the inside radius and mark it out as well.

Next you'll need to attach a bottom plate to the floor on the radius you marked. Cut out the bottom plate from a sheet of ¾-inch plywood. If you have room to lay down a full sheet directly over the area where the bottom plate will be installed, you can

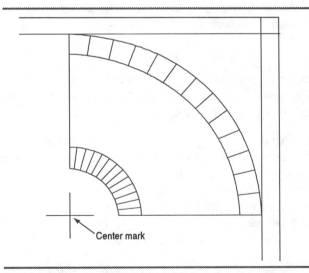

Figure 9-28   Circular stairs are laid out from a center mark. You'll pull the outside and inside radius from this same mark. The plans will give you the dimensions and you simply measure in from each perpendicular square stairwell wall and find the intersection point. Find the riser layout points by dividing up the lengths of each plate radius evenly.

Figure 9-29 Circular stairs are made up of individual rise and run assemblies. A radius plate must be cut and attached to the floor on either side of the stair. Then the plates are divided up into equal pie-shaped treads. Cut the supports on the outer wall before applying the tread. If you cut the treads properly, each stud will be plumb.

use the same nail in the floor that you used to detail the radius. Otherwise you'll have to set up a new nail in a more open area.

Again hook your tape on the nail and mark out the plywood as you walk backwards. This creates a perfect curve that matches your line on the floor. On the outside radius, subtract 3½ inches on your tape and make another mark so you scribe a 3½-inch bottom plate. For the inside bottom plate, add 3½ inches to the original radius.

Set the plywood on some sawhorses and cut out the radius plate. Then lay it on the floor and mark out the adjustments for each end cut. When it's perfect, use it as a pattern to make another one exactly like it. The two together will make a 1½-inch-thick bottom plate. If you're working on a slab, slide some roofing paper under the plywood before you nail it to the concrete to protect it from moisture.

Once the plates are down you'll need to lay them out. You'll have one less tread than the number of risers you calculated. The trick is to divide the length of the radius walls into even tread depths.

Let's say you have 16 risers, which gives you 15 runs. Measure the length of the bottom plate by bending your tape around the radius of the plate. Assume the outside plate measures 236 inches and the inside measures 93 inches. Now divide each of those measurements by 15:

$$236 \div 15 = 15.73$$

$$93 \div 15 = 6.2$$

Figure 9-30 Each supporting wall assembly in a circular stairway is one rise taller than the last. Figure the height of each rise the same way as for a straight stairway. Notice how the treads were left long over the sides to protect the drywall.

The treads will be 15¾ inches wide on the outside radius and 6¼ inches on the inside radius. Mark out a line every 15¾ inches along the length of the outside plate and every 6¼ inches along the inside plate. See Figure 9-28 (but there are only 14 risers illustrated there — don't let that throw you off!). You should end up with 15 exactly spaced markings on both the inside and outside plates. These lines indicate the face of each step's rise.

Closed stringers are actually small walls constructed separately for each rise and run. The walls can be stick framed separately, or attached to each progressively taller 2-by riser as shown in Figure 9-29. Nail the face of each wall or riser along the layout markings you put on the bottom plate. The 2-by riser supports the front of the tread material, and a 2 x 4 attached along the bottom of each 2-by riser supports the back of the tread. Look carefully at Figure 9-29. The first wall is the height of your calculated rise, *minus* the tread material. Each wall that follows is one rise taller than the one before it.

There are so many different ways of building radius walls for stairs that it would take a whole chapter just to cover them all. Usually they are done like the one in Figure 9-29, with the framer making any necessary changes in the design as he goes along. The trick is to get each rise and tread the same, both in rise height and tread shape.

On circular stairs with larger treads, you might choose to frame individual walls, and then joist out each tread with 2 x 4s. If the outside tread dimension

Figure 9-31   On large tract jobs, stair cutting gets into full production mode. Once a stairway is built for each unit, the numbers will never change. An apprentice carpenter can cut up and stack all the material for each stairway out in the yard. This would include all horses, risers, and treads. Notice the 1 x 6 is already attached to the side of the horse.

was fairly wide, the ¾-inch tread might flex a little over time. If you choose to go with this design, remember to frame your walls 3½ inches shorter to allow for the 2 x 4 joists.

When all the walls are built, cut and attach the treads. Run the grain on the plywood parallel with the width of the treads. Framers often leave the treads a little long on the outer edge as seen in Figure 9-30. This plywood lip protects the drywall.

The trick to attaching the treads is to get the first few walls plumb. As you work your way up the stairway, you want the walls to be plumb as you pull them into the back side of each tread. If you aren't careful about this, the last tread could be much larger or smaller than the other treads, depending on the direction the whole unit is leaning. Get the first few walls plumb and brace them off if needed. Then as you progress, keep checking that each wall

down the line falls pretty close to plumb. You can't vary the treads more than ⅜ inch in size, so you have very little room to play with.

Let's Talk Money

On a typical tract job, a set of stairs will pay from $7 to $10 a rise for an easy, straight set. You may get up to $500 for a complicated stairway. Expect an extra $30 to $70 per landing. Stair cutting pays better than other framing jobs. If you have the chance to cut stairs on a sizable tract, you can make good money while the job lasts.

Every apprentice carpenter should have the experience of working under a master stair cutter. Ideally, an apprentice would start out cutting packages of horses, risers and treads (like the ones in Figure 9-31), then move into building landings and assembling the stairs.

Ten

Shear Walls

S hear walls are structural walls engineered to help support a building during periods of high stress — earthquakes, high winds, and uneven settling of the soil. When the framed wall has been plumbed and lined, you'll sheath shear walls with structural plywood sheathing or waferboard, whichever is specified in the plans. The engineered specs give you the nailing schedule. In some areas it's the custom to sheath the entire exterior of the house with plywood or waferboard. In others, the structural page may only detail a few walls as shear walls.

Most large houses and commercial buildings include a network of interior and exterior shear walls that transfer stress from the roof and walls to the foundation. Interior shear walls will often be the only interior walls that are attached to the concrete with bolts. The engineers often specify heavy-duty hardware installed in a shear wall (like hold-downs and PA straps) to strengthen the connection from wall to floor, or first story wall to second story wall.

Framed walls higher than 14 feet (as in Figure 10-1) have virtually no resistance to shear stresses when they support a heavy roof. They're dangerously top-heavy. The drywall and stucco contribute some shear value, but not enough to withstand an earthquake or strong wind. Under that kind of pressure, the top-heavy walls can actually fold or flatten lengthwise if they don't have enough shear strength.

I remember seeing the devastation to a multistory hospital after a California earthquake in the early 1970s. The building was basically a ten-story square box, with two shallow rectangles attached on opposite sides. Apparently the two rectangles weren't anchored very well to the main structure and didn't have enough shear strength on their own. Both

Figure 10-1 A shear wall must extend from the bottom plate to the top plate to do its job.

plywood. If the wall is taller than the sheets, install a row of blocks at the break line. If the original detail person did a good job, all the studs will be laid out 16 inches on center and all the blocks will be in place.

If you're planning to hang shear panel for a piece price, check out two points before bidding the job. Do you have to install missing edge blocks for the horizontal breaks, or add studs where the plywood's edge falls (the vertical breaks)? If the vertical break falls on a stud, you're set. If not, you have two choices: you can cut down every sheet, or add a stud at the breaks. Either way, you should get a decent hourly wage for that extra work. On a poorly-framed wall, it will take you longer to prepare the wall for shear panels than it will to hang and nail the panels! Make sure you're not donating that extra time.

It only takes a few moments of poor stud layout to add hours of extra work for a person installing shear panel. That's why I spent so much time explaining shear wall layout in Chapter 4. A wall that doesn't have studs exactly 16 inches on center down its *entire* length is useless for a framer attaching plywood that's 48 inches wide. But a wall that's laid out properly — oh, what a blessing! What a money maker!

had peeled away from the main building and lay flat on the ground. In the part of the building that remained standing, you could count ten hallways that ended in the open air. Any time I begin to feel that the job I'm framing is over-engineered, I remember that hospital, with its gurneys dangling out of the open hallways ten stories up.

Preparing to Hang Shear Panel

The material most commonly used for shear panel is ³⁄₈- or ¹⁄₂-inch structural plywood. To maintain its strength, you have to nail all edges of the

■ Installing Hold-downs

You can't hang shear panels until all the hold-downs are installed. A hold-down is an angled piece of metal that's bolted to the side of a post and to a threaded steel stud bolt that's set in the concrete when it's poured. See Figure 10-2. When you install the shear panel, you'll nail the shear panel to this post (and all the other studs in the wall), transferring the strength of the post to the entire length of the wall. The hold-downs may be installed by hourly carpenters or included in your piece price for hanging the shear panels.

The structural or foundation plan will show the size and location of the hold-downs. They're usually installed at both ends of a shear wall, ensuring that the entire wall is solidly held to the slab. If you

Figure 10-2 Hold-downs must be installed before you hang the shear panel. Sometimes it's easier to predrill the post that the hold-down is attached to before it's installed. Notice in the photo that the stud bolt embedded in the concrete was set too low in the concrete. The nut couldn't fully engage the bolt to secure the bottom of the hold-down. I notched the bottom plate ½ inch so the hold-down could move a little farther down on the bolt.

have a hold-down on a post at either end of a wall, the only way the wall will ever move is if the concrete does (hopefully never!).

The structural page of the plans will also give you the size of the post (4 x 4, 4 x 6, or 6 x 6) built into the wall at the appropriate distance from the stud bolt in the slab. The bottom hole in the hold-down fits over this stud bolt and is fitted with a nut and a washer. The distance from the edge of the post (the vertical edge of the hold-down) to the center-line of the stud bolt will be anywhere from 1½ to 2¼ inches, depending on the size of the hold-down.

The manufacturer of the hold-down provides a spec sheet that identifies the size of the stud bolt, and the exact distance between the stud bolt and the post.

There are two problems you might face when you get ready to install your hold-downs. As we mentioned in Chapter 3, the threaded stud bolt may be placed in the concrete by someone who has no experience with installing hold-downs and doesn't understand the extremely small margin of error the job allows. The carpenter must make the hole in the hold-down's base go over this bolt, and at the same time fit snug against the side of the post it attaches to. If the post is installed near a window or a corner, the placement of the concrete-embedded bolt is critical because there's no room to shift the post around. There's only about ½ inch tolerance in the bolt-to-post measurement.

That's why some framers take the time to mark out the location of each hold-down for the concrete person *before* the concrete is poured — even if they don't get paid for it.

The second possible problem occurs if your post was installed in the wall when it was built. Then you're relying on the accuracy of the layout person marking the plates and the wall framer following those marks to install the post. Sometimes the post will be predrilled to accept the hold-down. Are the holes too high or too low? When the holes on the side of the hold-down line up with the drilled holes in the post, does the bottom plate of the hold-down lie flat on the wall's bottom plate?

If the holes aren't predrilled, can you maneuver the drill at the right angle to drill the holes without removing the post? You may find studs are laid out right next to the post, making it impossible to maneuver your drill. You may have to remove the stud or the post in order to get a good angle for your drilling.

Hanging the Shear

Hang shear panels with the grain running the same direction as the studs. On most exterior walls, you don't have to fill in above and below the windows or doors, unless your contractor insists. A window or door breaks up the connection from top to bottom plate that gives the shear its strength, so filling in the spaces above and below is a waste of time and material.

But there's an exception to that rule. When your layout results in a whole sheet covering the window or door opening, it's easier to just put up the sheet and then cut out the opening.

If the wall will be covered with stucco, it's not necessary to fill in above and below windows, as the stucco will hide this small change in wall thickness. They just apply the stucco a little thicker. If wood siding will be installed, the contractor may want you to make the wall above and below the windows and doors the same thickness as the shear panel so the wood siding will install cleanly. You can run the grain on the shear panel horizontally on these fill-in pieces to use up scrap.

■ Spacing the Shear Panels

Changes in moisture tend to make shear panels expand and contract after panels are installed. Even when a building is completed, it allows a certain amount of air to pass through the stucco and drywall into or out of the building. As air is drawn into the structure, it brings moisture in with it. In cold climates, a plastic moisture barrier should be installed under the drywall on the interior side of the wall. This barrier protects the drywall from moisture. But the shear panel isn't protected, and as temperatures rise and fall, and the interior of the building struggles to balance with the outside weather, the shear panel is subject to this regular cycle of higher and lower moisture levels.

As the shear panel absorbs moisture, it swells. As it dries out, it shrinks back to size. If you install the shear with each sheet tight up against the adjacent sheet, there's no room for swelling. Sooner or later, it'll buckle. This will eventually lead to cracks in the stucco. The solution to this problem is very simple. Leave a ⅛-inch gap between each sheet. That's all the room panels need to expand. An 8d nail is exactly the correct width for this. Just tack

one at the edge near the top of the panel you've just hung and another along the edge near the bottom. Then slide the next piece up tight against these nails. Once the sheet is in place, you can remove the nails and be assured that you have a perfect gap.

■ Installing the Shear

The first step to actually hanging the shear is to measure the length of your plywood sheets. If you're working on a single-story structure with typical 8- or 9-foot walls, the shear panel will go from the bottom plate to the double top plate. If it's a two-story structure, you want to go from the bottom plate to somewhere around the middle of the second story rim joist. This length lets you edge nail the bottom panel in both the top plate of the first story wall and the rim joist of the second story. That makes an excellent connection between the wall and the floor.

You'll need to get a horizontal row of nails into the top plate as well as the joist. Top and bottom plates *always* need a row of nails. If the shear panel continues on the second floor, install the upper sheet from the middle of the joist (on top of the first sheet you set). Again, edge nail both in the rim joist and the bottom plate of the second story wall.

If you're installing shear panels on a balloon or rake wall, the first row of panels should go from the bottom plate to the shear block in the wall. If the detail man installed this block correctly, the center of this shear block should be at either 8 or 10 feet from the floor. That way you can hang up full uncut sheets. Having to stop and cut sheets before you hang them really slows you down. Make sure the contractor hears about poorly set blocks, as well as misplaced studs that don't fall on perfect layout. If you're piecing out shear panel, you should be able to count on blocks and studs that are properly spaced in the wall.

After you hang the first row, jump up and measure the length of the top row of sheets. Then cut them to size all at once. See Figure 10-3. Be sure to stagger the joints vertically. If you have any doubt about what's expected, start asking questions.

Figure 10-3   When you've measured the length of the upper sheets, lay them out on a set of sawhorses and cut all panels to length at once. Before you hang any panels, check that the shear blocks are lined up to catch the nails from the top of the lower sheets and the bottom of the upper sheets.

Are the blocks set up for it? Does the price reflect it if they aren't? Remember, setting blocks from a ladder is a tremendous headache.

■ Hanging Panels on the Second Floor

Framers will often square up and shear the wall before it's raised so nobody has to climb up there and do it later. If not, you've got to get it up there somehow, hopefully on a calm, windless day! Hanging shear panels on the second floor or on tall rake walls can be quite an experience.

If you're working alone, the only way to get the sheet up is to have some kind of support along the bottom edge to hold it while you maneuver it in place and then nail it. One simple way is to set a few 16d nails into the joist along the top edge of the lower row of shear. If you set these nails at a slight upward angle, they should support the bottom of the panel while you nail it (Figure 10-4a).

There's one problem with this setup: not all sheets will hold perfectly flat for you. Some are warped, floppy, bent or windblown. As soon as you have it in place, you reach for a nail. That causes a shift in pressure somewhere on the sheet and it starts to shimmy. Before you can react, that sheet is heading down (Figure 10-4b). There isn't a whole lot you can do about this except try to anticipate the shimmy before it comes. Try setting some nails in the sheet before you put it in place. Then you can get it started before you have to reach for the nails.

Shearing tall walls before they go up is a great idea if the timing on the job allows. Then again, I've seen some real magicians working up high. Shearing tall walls isn't a job for everyone. You need patience and a willingness to learn how the sheet wants to move. You also might need a buddy. Sometimes there's no way to do the job alone. If the nails aren't doing any good, you're going to have to get someone to help you. Either that or cut the sheet down to a more workable size.

■ Hanging Shear Panels Around Windows

There are two ways to install shear panels on a wall that has windows. Some framers hang uncut sheets right over the rough framed window. Later they go back and cut out the window openings from the outside with a circular saw or from the inside with a reciprocating saw. Other framers measure each window and cut the sheets before hanging. This works especially well for hanging second story sheets. Once your window opening is cut out, you can hold the sheet in the center rather than along one edge. That makes it a lot easier to handle.

Figure 10-4a   It's hard to hang upper sheets when working alone. It looks like this framer has the sheet balanced on the lower shear. But when he reaches for a nail, oops . . .

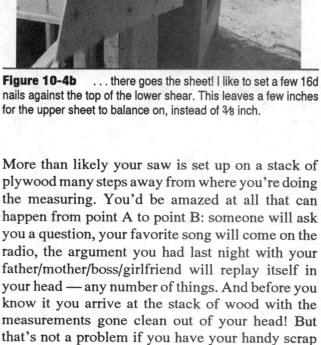

Figure 10-4b   . . . there goes the sheet! I like to set a few 16d nails against the top of the lower shear. This leaves a few inches for the upper sheet to balance on, instead of ⅜ inch.

Once I watched a framer go all the way around a building, measuring pieces and drawing a map of where they went around window and door openings. He cut all the panels, including panels with window and door openings. Then he hung all the panels. It looked screwy to me (he had quite a few repairs to make along the way) but I guess it worked for him. I prefer to hang all my full sheets first, then measure and cut the panels that surround windows and doors. Once your full sheets are up, you can measure off of them to your window openings and then transfer these numbers to the sheet you'll cut.

When you're taking measurements for any opening, try making a quick drawing on a scrap piece of wood and writing all your numbers on it.

More than likely your saw is set up on a stack of plywood many steps away from where you're doing the measuring. You'd be amazed at all that can happen from point A to point B: someone will ask you a question, your favorite song will come on the radio, the argument you had last night with your father/mother/boss/girlfriend will replay itself in your head — any number of things. And before you know it you arrive at the stack of wood with the measurements gone clean out of your head! But that's not a problem if you have your handy scrap of wood with all your measurements on it.

When you get to your stack of wood, transfer the numbers, chalk out the lines with your chalk box, and cut away. Try setting your saw blade to the exact depth of the plywood you're cutting so you

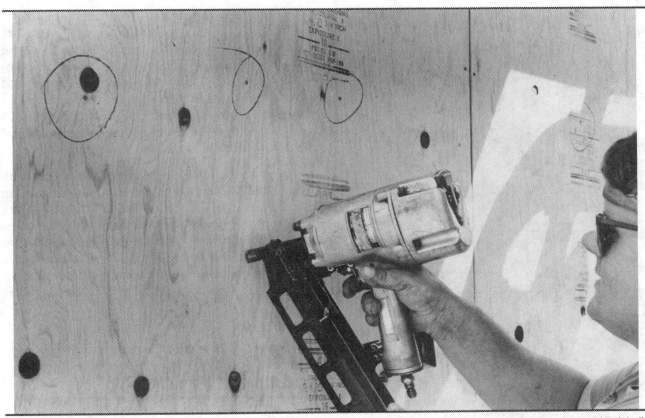

Figure 10-5 When nailing the shear, you have to imagine the stud behind the plywood. Listen to the nail as it goes in. A hollow sound means a miss and a solid thud means a hit. Notice the circled nails set in the middle of the studs. Set these when you hang the sheet. They make a good guide when finish nailing.

don't have to move the sheet off the stack to cut it. Better yet, set it just under the depth, then crack the skin that's left on the other side. If you set it too deep, you'll scar the sheet under it, sometimes rendering it useless. If you're working with a partner, try having one person set full sheets while the other comes behind cutting and filling in the blanks.

Nailing the Shear

As you hang the panels, mark on the face of each sheet the centerline of the studs that will support the panel. That makes it easy to know where to nail. It's not enough just to make a mark where you *think* the stud is. Always drive a nail through the plywood into a stud to be certain that there really is a stud there. Then mark the location by circling the nail. See the circled nails in Figure 10-5. Mark on the panel the location of all plumbing and electrical lines that you hide when hanging the panel. If you fail to warn the nailer of a hidden pipe or cable, be prepared to pay for the repair.

It takes a little practice to get the hang of nailing shear panels. You have to develop an eye and an ear — an eye for where the stud has to be and an ear for the difference between a hit and a miss. A hit makes a solid thud and will cause the stud to bounce a bit. A miss makes a flat sound that doesn't pull the shear tight to the stud. A few shiners (misses) are to be expected, but rows of them aren't acceptable.

Figure 10-6 Contractors will often hang the shear panel themselves and then sub out the actual nailing to someone set up with good nailing equipment. The nailer is then responsible for the inspection and must return if it doesn't pass.

The shear panel schedule in the plans will give you the spacing for the nails. It might call for 4 and 12 nailing, for example. That means all the edges of the shear panel (including any edges around windows and doors) would have a nail every 4 inches, with field nails every 12 inches.

A lot of contractors like to sub out the nailing to shear nailing specialists, like the one in Figure 10-6. There are two reasons for this decision. First, the nailers usually have a large trailer-mounted compressor that can handle endless lengths of hose and any number of guns. Although the contractor probably owns his own portable compressor that can run one gun, he knows he can get the job done in half the time with a couple of nailers working at once.

Second, building inspectors love to pick on shear panel nailing. Is it nailed at 4½ inches on center, instead of 4 inches, as the code requires? The inspector will notice the difference. Was the compressor turned up so high that nail heads sank too deeply? Mistakes like that catch any inspector's eye. When that happens, the contractor simply has the nailer do the work over, if he ever wants another nailing job.

What's the Pay?

Piecing out shear panel work will pay from $5 to $8 a sheet to hang and nail, depending on how many stories you're going up. I've seen jobs with high walls where the sheather and nailer could name their own price. There aren't many framers who can nail shear panels at 20 feet above the floor. It often takes rented machinery to get up to these heights. If a contractor is hanging his own shear, he's paying an hourly wage of $8 to $20 an hour, depending on who he has available. If the shear can't be reached from a ladder or placed from up high, he's also paying for scaffolding or a machine to get a framer up there. That's why a lot of people choose to shear second story walls before they raise them in place.

ELEVEN

Stacking a Truss Roof

In the old days, every roof member (rafter, ridge, valley, hip) was cut and placed individually. Many roofs are still cut this way, especially really complicated ones on custom homes. But truss roof packages are being used more and more. A truss roof takes half the time to install and reduces the contractor's labor cost.

Every roof truss is custom designed, engineered, and manufactured for the house it will go on. These labor-saving trusses are used on many residential tracts and custom homes. Roof trusses are shaped like a triangle, with two sides acting as rafters and the lower chord acting as the ceiling joist. When you raise a truss in place, you are installing the ceiling joists and roof at the same time. See Figure 11-1.

Another advantage of trusses is that the spans can be longer. Web members installed between the upper and lower portions give added strength. On most every truss-designed roof, the trusses should only bear on the two exterior walls. A truss should almost never need an interior wall for support. That gives the designer more options when laying out rooms in a house. And the actual framing members can be made from smaller lumber.

I've noticed, however, that truss roofs tend to sag and show signs of fatigue long before a conventionally-stacked roof will. Still, there are so many advantages to a truss roof that it seems eventually every home will be built that way. Part of the problem may lie in the fact that most trusses are designed for spacing 24 inches on center. When the ½-inch roof sheathing is applied, the weight of the final roofing material sometimes causes the plywood to dip a little over time. Although it's approved by

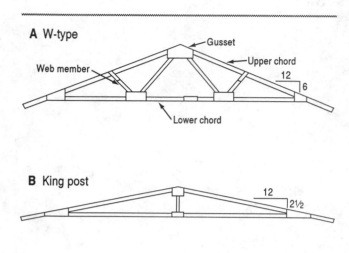

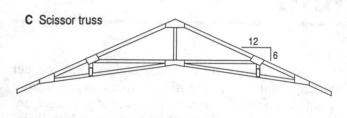

Figure 11-1 Trusses come in three basic designs. The W-type have inner webbing and are used for pitches 3 in 12 and up. King post trusses are for very shallow pitches. The scissor truss has a pitched ceiling joist chord for vaulted ceilings.

the codes, and almost universally done this way, most experienced builders will agree that truss roofs are not as long-lasting as conventionally framed roofs.

The truss is commonly made up of 2 x 4s held together with plywood or metal gussets. Each truss is so light that one person could probably spread trusses and stack a whole roof alone. But be careful when handling trusses. Don't let the truss bend too far while it's laying flat. They're designed to take considerable vertical loads but will break easily if bent too far horizontally. That's why a truss stacking crew should include at least two craftsmen.

Stacking conventionally cut roofs is a job that's usually reserved for experienced carpenters. Roof framing can be complex. But using roof trusses is comparatively easy. If you've never done any roof

framing at all, trusses are a good place to begin. Although a truss package may look complicated when it's delivered, it's really a simple system to organize once you've installed a few.

After beginners get over being intimidated by the scope of the work, some still shudder at the thought of walking on a 3-inch wide wall, 20 feet up, dragging a hundred-pound web of wood. It takes some framers years to get their "legs" — and some never do. Walking top plates has nothing to do with physical coordination or strength; it has everything to do with *mental* coordination and strength. It's a fear that each framer has to confront. Either you overcome it or you give up walking the plates.

Until you've done it, you'll never know if you can walk walls well enough to stack. It's not a job for everyone, although I've known carpenters who seem to thrive on danger.

Here's the trick to walking walls: Never look down, and never focus past the top plate. Keep your eyes focused ahead of you. Learn to trust your feet; they're smarter than you think.

Spreading and Stacking Trusses

Trusses are delivered banded together in separate packages of about ten to twenty trusses. They're banded in a logical sequence, with a gable truss on the end and any special trusses organized in the order you'll use them. I recommend leaving them banded until they've been placed on top of the walls with a pettibone or crane. But first, you've got to identify what goes where.

Truss packages are usually organized by rooms. Sometimes you'll have a longer "tail" or overhang on one side than the other. Avoid laying a package on the wall upside down or with an overhang tailpiece on the wrong side of the building. That would require spinning each truss before you can roll it. Except for dragging in a straight line, trusses are very hard to move once the package is up on the

Figure 11-2 Getting the truss packages up on the wall is easy with a crane or pettibone. Make sure you have the proper overhang tail on the right side so you don't have to spin the trusses when you spread them!

walls. Either lay the trusses flat on the walls with the ridge points facing out (see Figure 11-2) or leave them standing as close to the exterior wall as possible. If they're standing, install a few temporary braces on the back side of the package before you break the bands.

As long as you have the lift available, place the small fill packages and the blocks upstairs. Try to get all the trusses laid out in approximately the correct location before rolling any up.

Start spreading the main body of trusses. Have one person work either end of the truss. Grab the truss by the tail and walk steadily along the top plate, dragging the pointed top end of the truss behind you, until you reach the other end of the

building. Now flip the truss over so that the ridge is pointing out of the building. This will support the first truss, much like the package is supported in Figure 11-2. The next truss ridge will then rest on this one after it's flipped and so on with each following truss.

Why don't we set up the whole package with the ridges pointing in the right direction, and thus avoid flipping each one? Because there's no way to carry this flimsy beast without dragging the point (ridge) behind you. Continue dragging and flipping the trusses one at a time, layering them like a deck of spread-out cards. See Figures 11-3 and 11-4. Place each truss 2 feet closer to the main pile. That way each truss should be very near where you'll roll it up.

Figure 11-3   Spread the trusses so the end at the wall plate is in approximately the correct position. If the trusses will support your weight, walk out to the ridge end and nail a ridge block to one side of each truss. In areas like garages where there are no interior walls, the trusses won't support your weight. So you'll have to block the ridge after the trusses are rolled up.

When you get to the end there should only be a few trusses left to flip and stack in a pile. That gives you a little working room for the first couple of trusses.

■ Preparing the Blocks

Once trusses are spread out, get your blocks ready. Each truss requires a 22½-inch block over the top wall plate on each end and one at the ridge. One 22½-inch block plus the thickness of the truss (1½ inches) keeps the trusses 24 inches on center.

It speeds the work if blocks are nearby when you're placing trusses. You should have been delivered enough precut blocks to do the entire roof. If not, you'll have to cut blocks. After they're cut, I like to set a toenail in the edge of each. Then I hang the blocks on a long 1 x 6 I've tacked up near the

top of the wall. Or use a single 16d nail to hang each block as shown in Figure 11-5. As each truss is rolled, I just bend down, grab the nearest block, and nail it in place. This saves carrying around a handful of blocks. Cut blocks for use at the ridge at the same time you're cutting blocks for the truss ends. Also lean a few lengths of 2 x 4 against the wall for special blocks and bracing.

If you're piecing out the truss stacking, you're not required to cut your own 22½-inch blocks. These should always be provided! You *will* have to cut any special blocks.

On tall roofs, I like to tack all my ridge blocks to the trusses while they're still lying down. You can see these ridge blocks in place in Figure 11-3. It's a lot easier to have the ridge block nailed to at least one of the trusses when it's rolled up. Of course, you still have to climb up there later and nail

Figure 11-4   Roll up the first truss and install a temporary brace to hold it in place. Notice that the tail on this truss has been cut off flush with the wall line. That's because the tail falls in a valley and no projection is needed. The tail on the next truss will be cut a little long, where the tail of the perpendicular truss hits it. When in doubt, leave plenty of rafter tail. That simplifies work for the tradesman who installs the fascia.

the other end of the block to its neighbor truss. But at least you don't have to carry them while you climb up to the ridge.

Those blocks also help keep heavy trusses balanced in place while you tip up the rest of the trusses. Then you only need a vertical brace to support the first gable truss. The rest will kind of float until you jump up and back-nail the ridge blocks.

Once all the trusses are spread out flat, walk out to the ridge. Nail a block to each ridge. Be careful about walking on the body of any truss spread out flat unless there's an interior wall supporting your weight. Try to nail all the ridge blocks you can while the trusses are still down.

I'll admit that some framers never install trusses this way. Some block the ridge as they go, using three framers; one on each exterior wall, and another to work the ridge. Others lay a flat 1 x 6 along the rafter member of the truss, as high up towards

Figure 11-5   Cut, nail and hang your blocks where they're handy when trusses are rolled up. Just reach down and grab a block when you need it. These trusses were rolled by one person from the opposite wall; that's why they're standing, but not blocked. They have yet to be nailed in place. It's always best to nail them in place as you block, rather than nailing them down and then trying to squeeze a block in later.

Figure 11-6   A gable end truss will look different than the rest of the trusses. Notice the 16-inch O.C. studs built into this gable end truss. This allows for the attachment of stucco wire or wood siding.

the ridge as they can reach, and tack a temporary nail into the truss to hold it in place until they jump up and do the ridge blocks.

There's really no right or wrong way to install roof trusses. But I hope this chapter describes one good way to get the job done. Then watch an experienced framer set roof trusses and you'll pick up a few new tricks. Just a few minutes of observing on the job can teach you more than hours of schooling.

Rolling the Trusses

To begin rolling, start by bracing up the first truss. On a gable roof, the first truss is naturally known as the (you guessed it) gable truss. This truss stands flush with the outside wall. It's different from the other trusses because it includes vertical wall studs every 16 or 24 inches on center. See Figure 11-6. This serves as backing for the stucco or siding.

■ Rolling the Gable Truss

With one person on either end of the truss, roll it up into place. Check to see that the truss is centered between the supporting exterior walls, not hanging too far over either wall. The bottom chord (or ceiling joist section of the truss) should be the exact length of your measurement from the outside of one supporting exterior wall to the other. Nail the bottom chord of the truss to the top plate of the wall once you have each end lined up.

The gable truss should also line up with the outside edge of the exterior wall under it. Use 16d nails every 16 inches toenailed from the bottom chord down into the top plate of this exterior wall. When the gable truss is in place, lay a long 2 x 4 diagonally up the wall and beside the truss. Tack it to the truss and a stud to stabilize the gable truss and keep it in line with the plane of the wall below.

Now roll up the next truss until the block at the ridge meets the ridge of the gable truss. Again, make sure that the truss is centered between the supporting walls before you do any nailing. If for some reason some bottom chords are a little too long or short, keep them all flush with one supporting wall. This keeps the ridge line of the trusses as uniform and straight as possible. If the walls aren't perfectly parallel with one another, one side won't be flush after a while. Still, it's best to always match up on one side and let the other land where it may.

Before you put the side blocks on, find out whether the block should be set vertically and flush with the outside wall, or positioned as a frieze block. Frieze blocks run square with the truss, with the back side of the block just touching the wall. The block in the foreground in Figure 11-7 is a frieze block. This block gives the stucco or siding something to finish against. In Figure 11-6 you can see the finished line of frieze blocks in place along the front of this garage wall.

Every third or fourth block will usually include a screened vent. These vent blocks should be shown on the plans and will usually be supplied by the truss company. Figure 11-5 shows a vent block waiting to go in.

Some of you might be wondering about something at this point: How do you know where to nail the truss if there are no layout marks on the top plate? Good question! After you raise the gable truss, install the first 22½-inch vertical or frieze block. When you pull up the next truss and slide it against this block, it will be exactly 24 inches on center from the gable truss. With each truss and block you install, the layout will con-

Figure 11-7   Don't roll up a truss until you've installed the spacer block that marks the location for that truss. The first two blocks here are flush with the exterior wall. The rest of the blocks will be frieze blocks. When the fascia person cuts in the valley, he'll also cut in these last two frieze blocks.

tinue as long as you use a 22½-inch block, and you install it tightly, with no gaps. The precut block makes the layout. You only have to worry about keeping the bottom chord centered over the two exterior walls.

Figure 11-8   Pull up the next truss and nail it to the blocks you've set. Also get a couple of toenails down into the top plate of the wall. Notice that this gable truss is not along an exterior wall. This truss landed on the wall that separated the garage from the house, which in some jurisdictions needs a one-hour fire protection barrier. This is provided by drywalling the truss up to the underside of the roof sheathing. The gable studs will help to attach the drywall.

Figure 11-9   Toenail the truss down to the top plate of the wall before setting the next block.

■ Rolling the Other Trusses

When the gable truss is up, nail a block over each wall or set it square as a frieze block. Now roll up another truss (Figure 11-8) and slide it into the first block you installed. Put a couple of toenails down through the truss into the top plate of the wall, as in Figure 11-9. Place the free end of the block so that the back edge is tight against the wall and then face nail through the second truss into the block. At this point the trusses should be pretty stable. Install another block and roll up the next truss. Continue until you've rolled up all the trusses.

If the building length isn't evenly divisible by 2 feet, the last gable truss will need blocks cut to a special length. When all trusses are up, nail in the ridge blocks (see Figure 11-10). If you installed them to one truss before you raised it, back-nail the loose end into its neighbor truss. Make sure all the ridge blocks are nailed tightly, or your trusses will start to tip slightly. If you left a 1/16-inch gap in each ridge block, and your wall blocks were all tight, the sixteenth truss would be leaning 1 inch. I guarantee that you will be sent back up there later to fix any loose blocks that cause the trusses to lean.

Once you're finished tightening up all the ridge blocks, sight down each gable end. Make sure the gable truss is flush with the exterior wall below. If you aren't experienced in sighting the plane of the gable truss with the outside of the wall plane, get a level and make sure the gable truss is nice and plumb. If not, remove the ridge block and cut another block to adjust the spacing between the gable truss and the next truss. The last gable block often needs a little adjusting to make the end of the roof structure level.

Rolling Hip Trusses

A hip truss roof looks more complicated lying on the ground, but isn't really much harder to master than a gable truss roof. Basically, you'll have a hip package on either end, with a row of conventional trusses in between. See Figure 11-11. The conven-

Figure 11-10   Make sure the first truss is braced well before walking the ridge to nail the ridge blocks.

tional trusses terminate approximately half the width of the building back from the outside wall that the trusses run parallel to. The set of jack rafters, which run perpendicular to the main body of conventional trusses, have ceiling joists built on them. When rolled up, these ceiling joists will nail up to the first conventional truss, which is often a doubler.

The first step is to establish the position of the first and last conventional trusses. Measure the length of the ceiling joist leg that's attached to the jack rafters which run perpendicular to the main body of joists. Because this is the length from the outside wall to the first conventional truss, it gives you the distance to place the first conventional truss from the outside wall. Then roll up the center package of trusses exactly as you would roll any conventional trusses. But make sure the first and last conventional trusses (which are probably doublers) are set at exactly the correct position.

The next step is rolling the hips. These will run from the center of the ridge on the first conventional truss down to the building corners. They should be provided in your truss package. Jack rafters fill in

Figure 11-11   This is a scissor hip assembly. The conventional scissor trusses at the center of the room were rolled up first. Doubler trusses were installed on each end of the conventional trusses. These doublers support the hip trusses and the hips support the jacks. What looks like a confusing mess from the ground is really quite simple to install.

along the hip. These jacks should also be spaced with blocks cut to 22½ inches long. Be careful not to bend the hip when you attach each jack. Sight down it now and then to be certain it isn't getting bent by the pressure of the jack rafters. Install the jacks from the top plate to the hip, with the ceiling joists attaching to the lower chord of the hip member.

Filling, Understacking and Ceiling Backing

Once the trusses are all rolled, you'll need to take care of any California fill, understacking, and ceiling backing that's necessary. We'll look at these operations one at a time.

■ Framing a California Fill

Occasionally you'll have one roof plane or surface that merges with another. See Figure 11-12. Obviously we couldn't leave the roof like this, as it would create severe drainage problems. We have to fill in the roof surface on the left with hip and jack rafters that run smoothly with the plane of the roof surface on the right. This is called a California fill.

With a conventionally framed roof, you'd have to cut separate valleys and jack rafters for a section of roof like this. But with trusses, you just frame one roof right over the top of the other. The trusses on the left in Figure 11-12 will support the roof that is built on it and planes in with the trusses on the right.

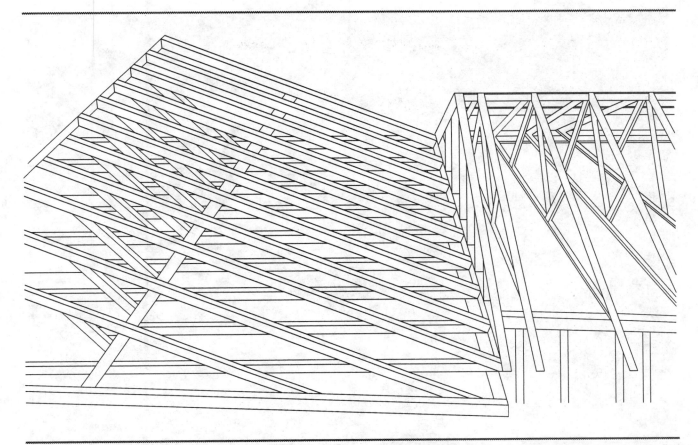

Figure 11-12   The ridge on the right has to be extended to intersect the roof surface on the left. The valleys that result are called *California fill*. This work may be done either by the roof framing crew or the sheathing crew. Be sure you know which crew has the responsibility.

In the end you'll have two valleys with the appearance of a conventionally framed roof, but made in half the time.

The first step in framing a California fill is to sheath and nail the lower roof surface. Then pull a string line along the top of the ridge of the standing trusses on the right, to find the point where the extended ridge line from those trusses will intersect the sheathing that will support it. I pulled a string line horizontally along the ridge of the roof on the right until the string intersected the sheathed plane of the roof on the left. (You'll notice that the sheathing's not on yet in the drawing!) The point of intersection is the top of the two valleys that will form the California fill.

Next, I located the position for the bottom of these two valleys. I pulled a string line parallel to the wall plate and along the tops of the trusses on the right. Where this line intersects the roof plane on the left, I made a mark. See Figure 11-13. Then I snapped a line from the first mark (at the top of the valley) to the mark at the bottom. The valley board runs parallel to this line, but an inch or so below it, to allow room for the sheathing to plane into the line.

Here's an easy way to find out how far the valley board should be below the snapped line in order for the sheathing to plane exactly. Run a string along the tops of the trusses, then slide the valley board up the sheathing until it just touches the string. Then nail it in place.

Figure 11-13   Begin a California fill by sheathing the roof under the fill. Use a horizontal string line to find where the two roof lines should intersect. That marks the position of the valley. Lay a valley board just below the line of intersection.

Now measure from the ridge of the last truss to the intersection point on the sheathing at the top of the two valleys. This is the length of the ridge. Cut the end of this ridge board that rests on the valleys with the complementary angle of the roof slope. The complement of any angle, when added to the angle, will give you 90 degrees. For example, the angle for a roof that is 5 in 12 is 22½ degrees. And 90 minus 22½ equals 67½. So the complementary angle of a 5 in 12 roof is 67½ degrees, which you can quickly mark out with your speed square.

Nail the California fill ridge in place (Figure 11-14). Sight down this ridge board to make sure it's in line with the ridge of the trusses. Nail the end with the complementary angle down into the 1-by or 2-by valley boards you nailed into the sheathing.

Then cut and nail valley jack rafters (Figure 11-15), trying to continue the 16- or 24-inch layout from the trusses.

It's standard practice for the truss company to supply these jack rafters. If they don't land at 24 or 16 inches on center from the first full truss, it really isn't your problem. You just put them in however they fit square with the ridge. That is, unless you're going to be doing the sheathing — or someone is paying you extra to fix what the truss company sent. Then you might want a clean layout. In the long run it's probably easier to put them where they land and then deal with the screwy layout when you're sheathing. When you're sheathing a roof, valleys slow you down real fast. You end up piecing most of it in anyway, so the irregular layout isn't such a big deal.

Figure 11-14   The ridge line in the California fill should be level with the truss ridge. Make sure it's in line as well.

Figure 11-15   The truss company usually supplies valley jacks cut to fit on 24-inch centers. Nail up the first jack square with the ridge and then lay out the rest from that jack. I might lay a second valley board here because one 2 x 4 isn't thick enough to support the full length of the complementary angle cut on the jacks.

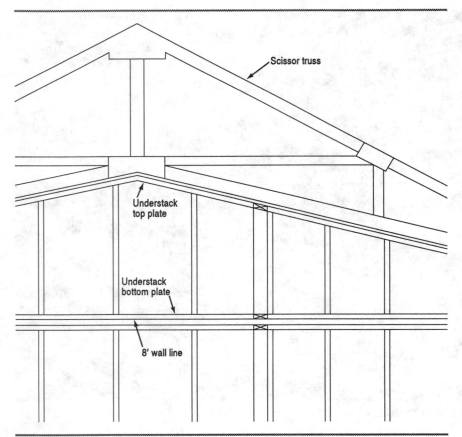

Figure 11-16   Understacking is the art of extending a wall to the underside of a raised rafter or truss. The understacking shown is built up to a raised scissor truss. Always match the stud layout on the original wall.

■ Understacking

Understacking is the process of building small walls that sit on top of the normal walls and extend up to a vaulted ceiling or the bottom chord of a scissor truss (Figure 11-16). The wall divides a room with a vaulted ceiling from another room with a flat one. The walls can be small uniform walls, or raked at an angle.

If the contractor is well organized when building the main walls, there will be little or no understacking to do after the roof is stacked. Balloon walls that divide rooms with different ceiling heights will be calculated and built from floor to the underside of the rafter or truss, eliminating any need for understacking. Similarly, rake walls will have been located and framed in their appropriate places.

On custom homes, however, a contractor who's inexperienced in dealing with balloons and rakes must resort to understacking to finish off the regular walls. Usually you'll see this on interior nonstructural walls. If the wall height was miscalculated on a bearing wall, the contractor's got big problems. By adding understacking on top of a wall, you create a hinge in the wall. For interior partition walls that only support the drywall, a hinge isn't a problem, as long as it's built straight! But for structural walls a hinge is not allowed.

When you install understacking, all you're doing is building short walls. You'll need a top plate, and if you want you can use a bottom plate. A bottom plate isn't required because the bottom of your studs can toenail directly into the top plate of the wall you're extending. Some framers like to build their understacking with a top and bottom plate, only because it's easier to deal with the little wall once it's built. Without a bottom plate, the studs tend to flop around when you're installing it. Either way is structurally sound.

The first step to understacking is to establish the height of the short wall. If you're filling to the bottom side of a flat ceiling, just take a measurement from the bottom of the ceiling joist to the top of the existing wall. Then subtract 3 inches for the top and bottom plate (or 1½ inches for just the top plate) to find the stud length. I usually take another ⅛ inch off the stud length just so I don't have to struggle with beating the wall into place.

The length of each plate will be the same length as the top plate of the wall you're stacking on.

Before you nail the wall together, there's one very important consideration you need to take into account. When laying out your studs, always match the stud layout in the wall you're extending. This helps keep things uniform for drywall or siding.

Once you have a layout, bang the wall together, then raise it in place. When you're satisfied that it fits, nail the bottom plate into the top of the wall you're extending. Make sure the bottom plate of your understacking is flush along the edges and on each end of the top plate on the wall you're extending. Or if you've omitted the bottom plate, toenail the studs to the top plate of the existing wall.

Once the bottom is nailed, use a straightedge or level to match the vertical plane of the understacking with the vertical plane of the wall. This is very important! If the main body of the wall is a little out of level, and you install your 3-foot piece of understacking at perfectly level, you're going to have one sorry looking mess once it all gets drywalled. I'm sure we've all seen the line in the finished, painted drywall that starts at about 8 feet where the top part of the wall is on a different plane than the lower part. Finished drywall tells more about the skills of a framer than almost anything.

If your lower wall is drastically out of plumb, you might consider doing some repair work to straighten it out. If it's just a little out, match your understacking with it, or split the difference between the little bit it's out and what the level reads.

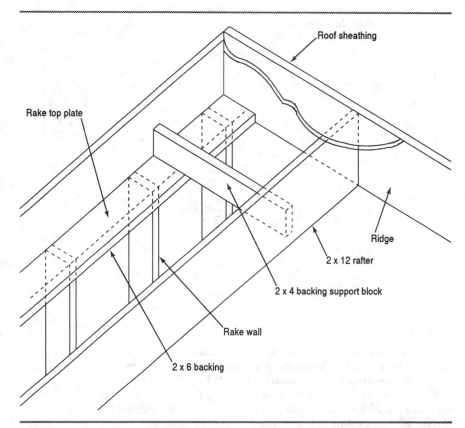

Figure 11-17 When installing rake understacking, always put the ceiling backing in first. A backing support block installed every 6 feet or so will help keep the backing level and sturdy.

Rake understacking is similar to straight understacking with a few exceptions. First, you'll probably want to stick frame your wall in place rather than building it on the ground. It's much easier and faster.

To begin a rake understack, you need to make sure you have backing in the ceiling (Figure 11-17). Always put your backing in first! Rake understacking almost always runs parallel to the rafters or bottom chord of the scissor truss. By installing a 3½-inch top plate, you create an area that will need ceiling backing, just like you did in joisting.

Once the backing is installed, locate where the top plate will go. Again, use a straightedge or long level to match the plane of the wall you're extending. Make a mark on the bottom (or shorter) side of

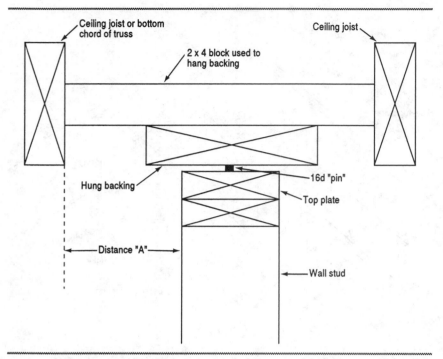

Figure 11-18   When installing ceiling backing on a truss roof, always float the backing between two trusses. Attach the backing to 22½-inch blocks and install it level with the underside of the trusses. When attaching the backing to the float blocks, start with distance "A" and subtract 1¼ inch or so. That's where you'll nail the backing to the blocks. "Pin" the backing to the walls with 16d nails every 3 feet.

To find the angle to cut on the top of each rake stud, you'll need to know the pitch of the ceiling you're building up to. If it's 5 in 12, cut the rake stud with a 22½ degree angle along its edge. When all the studs are cut, toenail the bottom of each into the top plate of the lower wall, and then nail the top of the stud into the top plate you installed. After you've installed a few, get down off your ladder and have a look at them. Are they standing straight? If they start to fall out of plumb, that means your progression was figured wrong, or the layout on the lower wall has changed. Make the necessary adjustments before you continue.

Ceiling Backing, Clips and Catwalks

For any wall framing from the room below that runs parallel to the bottom of the roof trusses, you need to install ceiling backing between the trusses.

It's important to "hang" or "float" all the backing on the trusses so it stays level with the bottom chord of the truss, no matter how that bottom chord moves (Figure 11-18). Trusses are built with a ½- to ¾-inch camber in their bottom chord so they don't rest on any interior nonbearing walls. Once the roof is loaded with tile or composition shingles, the trusses flatten out and settle. You want the backing to move with the truss to reduce stress cracking of drywall at the ceiling line.

Figure 11-19 shows a decorative column that extends to the ceiling line. Backing placed between the bottom chords of the truss is needed to join the column to the ceiling, and to provide backing for the drywall to nail to around the column. First cut the backing to length, then nail it to standard 22½-inch blocks. See Figure 11-20. Before doing any

the rake and another at the top. Then transfer these marks with a straightedge running up the vertical plane of the wall directly onto the ceiling backing. Once you've found the two end points, snap a line from one to the other. This is where your top plate will go.

Measure the length of the top plate, and nail it alongside your snapped line. Now, using your level, transfer the layout from the studs in the lower wall to the top plate of the rake wall you've nailed to the ceiling backing. Once you have a few marked out, check your measurements. There should be a common progression. For instance, if the first three studs measure 3 inches, 9¼ inches, and 15½ inches to each long point, each stud is growing by 6¼ inches. If the layout on the lower wall is a consistent 16 inches on center, there's no need to lay out the rest of the rake studs. Just add 6¼ inches to each subsequent stud you cut.

Figure 11-19 Walls and columns need ceiling backing for the drywall. For truss roofs, you'll have to float the backing.

Figure 11-20 Make sure the backing is flush with the bottom of the truss and attached to the truss so it will move with the truss. Notice the gap between the backing and the top of the wall. Drive pin nails down through the backing into the top of the wall to hold the wall in place when the braces are removed.

nailing, check the pieces you've cut with a straight-edge along the bottom of the trusses to be sure the backing is level with the underside of the lower chord. If it's perfect, nail these blocks between the bays of the trusses.

Always drive a few nails into the top of the wall, pinning the backing to the wall. But don't drive these nails down flush so the backing gets sucked down to wall level.

Chapter 7 on joisting described how to cut and install ceiling backing. Always work on the floor to take measurements and cut the backing, then hang it on a joist so you can grab it easily when you go up high. The only difference is that you aren't nailing the backing directly to the top plate of the wall. You're hanging it with standard 22½-inch blocks nailed to the top of the backing so it floats with the joists. Then "pin" the backing with 16d nails down into the top plate of the wall. But again, don't suck it down tight. When the trusses sink with the roof load, the backing will slide down these nails as it needs to. These pins are needed to hold each interior wall plumb when the temporary plumb and line joists are removed.

After you cut the backing, you will need to establish exactly where to nail your standard block. With an 8-foot piece of backing you'll need a couple of these blocks. But where along the 22½ inches do you nail the backing? This will change with each wall you encounter. Look at Figure 11-18 again. To find out where to nail the backing along the block, measure to find distance "A." Once you've found this number, transfer it to the block.

Notice how the backing extends a little over this measurement. The measurement is from the ceiling joist to the wall. But we want the backing to extend past the wall. So subtract 1¼ inches or so from distance "A" and that's where you nail the block to the backing.

The next step is to install truss clips. These L-shaped clips attach the bottom truss chord to the top plate of any wall that runs under and perpendicular to that chord (Figure 11-21). These clips hold the bottom chord on layout but don't support the truss. The clips are slotted on one side to allow the nail

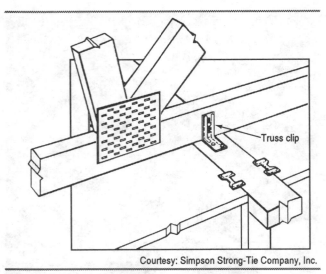

Courtesy: Simpson Strong-Tie Company, Inc.

Figure 11-21   A truss clip is attached to each truss and down into any perpendicular top plate. This holds the truss on layout. The nail that goes into the truss should be at the top of the slot in the clip to allow for the settings of the truss.

sent into the truss to slide as the truss settles. Nail the clip as high as you can on the bottom chord to allow for this movement.

You'll also need to run catwalks (Figure 11-22) and diagonal bracing in the attic space. A catwalk is a 1 x 6 laid flat on top of the bottom chord of the trusses. A catwalk should run the whole length of the building, from gable to gable, for each package of trusses. On a hip roof, you need only one catwalk on the center package. The catwalk helps to tie the whole structure together once all wall braces are removed. Also run a diagonal sway brace off of each gable end. Use a 2 x 4 installed at a 45-degree angle from each gable ridge down to the catwalk.

If you have exterior shear walls, you need a good connection between the truss wall block and the top plate of the wall. This transfers all the shear

Figure 11-22   The truss calculations on this house called for two 1 x 6 catwalks to hold the lower chords together. These catwalks had to be installed on either side of the middle support in the truss. Notice the backing floated along the wall.

value of the wall directly to the roof. The most common method is to use an A35 attached with 1¼-inch hanger nails on the backside of the block, as shown in Figure 11-23.

Here's a good trick. Attach the A35 to the block before you put it in place. Figure out which walls are shear walls before you begin setting your blocks in place. It's much easier to get at least half the A35 nailed while you're still on the ground. Later, after the block is installed, you can come back and nail the other half of the clip to the top plate of the shear wall.

You may be required to install hardware commonly referred to as *hurricane ties* (Figure 11-24). These attach to the side of the truss and twist in order to attach to the front of the exterior wall that the truss rests on. They're attached with 1¼-inch hanger nails, and go right behind the frieze block. They're extra insurance that the whole roof doesn't pull up off the top plate in high winds.

Your Pay for Stacking Trusses

A truss roof is probably the easiest roof to frame and a good place to start if you want to learn roof cutting. Even on a roof truss job, there are plenty of chances to develop and use your roof framing skills.

Roof framing tends to pay very well. The labor to raise a package of trusses is minimal compared to raising walls, if you can learn to walk walls! You can expect from 15 to 30 cents a square foot of floor space. These numbers change according to the economy or the amount of work involved.

Figure 11-23   On a shear wall use a framing clip to transfer shear forces from the wall to the roof through the blocks. A35 framing clips are very commonly used for this. Drive them onto your blocks before you install the blocks.

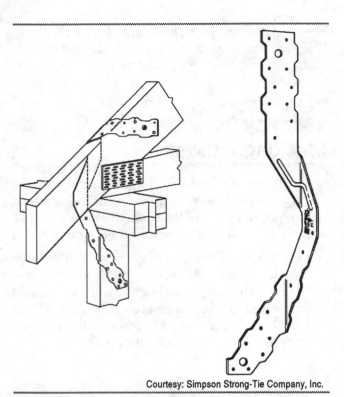

Courtesy: Simpson Strong-Tie Company, Inc.

Figure 11-24   Hurricane ties help to keep the truss attached to the wall in the event of high winds.

Always ask the general contractor if you're expected to do the California fills. Sometimes the sheathers take care of these. I've also seen military and commercial jobs where no ceiling backing was needed. The drywallers used special clips that eliminate the need for ceiling backing. If you don't have to place any backing, there's a substantial savings in time and material. Installing backing on the ceiling is usually as much work as erecting the trusses.

Always be certain of the understacking that's required before you start. Don't be the victim of an unscrupulous contractor who plans to drop a pile of extra work in your lap just as you finish the job. Eliminate all the surprises before you begin. Discuss what's expected until there's a clear understanding of the job scope. Settle the potential conflicts before they become arguments.

Piecework carpentry is seldom done under a written contract. If there's a problem, getting mad doesn't help, tempting though it may be. Find some way to negotiate a solution. You're more apt to get what's owed if you stay on good terms with the general contractor.

TWELVE

Cutting Conventional Roofs

When a house is designed with a truss roof system, a big burden is shifted from the framing contractor to the company providing the trusses. The truss company has calculated and cut the entire roof, right down to the frieze blocks.

With a conventional roof, the framer has to calculate and cut every common, hip, valley, jack rafter, ridge and block in the entire structure. It takes a healthy dose of experience — and confidence — to cut the entire roof package on the ground and be assured that it will work when you put it together. Although contractors generally use their most experienced and reliable carpenters to cut conventional roofs, it's never too early to learn the theories and tricks of roof cutting.

This is the age of the calculator versus the rafter length book. For rookie carpenters, rafter length charts are a valuable tool for double-checking their calculations. When you're starting out, your only goal is to get a roof properly cut and standing straight. If relying on rafter charts helps, that's fine. But after one roof, the book becomes a mental crutch that will slow you down. You won't find any rafter tables in this book. I find that after cutting a few roofs with a calculator, most carpenters' skill surpasses the book. When they understand how easy it is to figure out those numbers, they don't waste their time walking to the truck to get out the book. After I learned how easy it was to figure rafters, I never looked at another rafter chart.

If you want to cut complicated roofs, you've got to learn to calculate rafter lengths. I'll explain, in detail, the simple art of using a cheap calculator, the rafter framing square, and the width of your building to establish rafter lengths, including commons, hips, valleys and jacks.

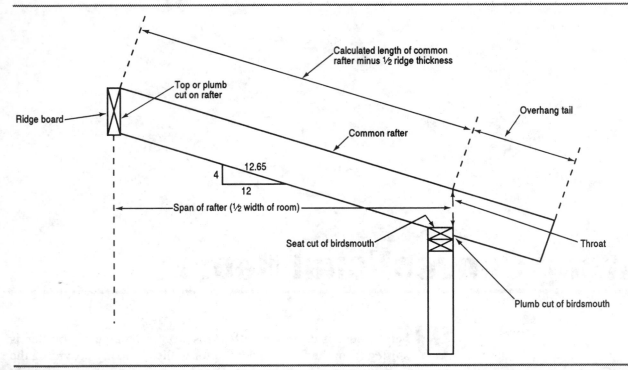

Figure 12-1 Here's a common rafter with all its features labeled. The rafter spans half the width of the room. This rafter is designed with the top of its plumb cut in line with the top of the ridge. The 4 in 12 triangle in the center of the rafter shows the relationship between the 4, the 12, and the length of common per foot of run (12.65).

Roof Terminology

Before we start doing any cutting, let's make sure you have a clear understanding of all the parts that make up the roof. The pitch of a roof is usually expressed in terms of the distance it rises vertically for every 12 inches it travels horizontally. In other words, a roof that was built at a 45-degree angle wouldn't be called a "45" or ½ pitch, as it once was. Today it's typically called a 12 in 12. It just so happens that the roof ends up at a 45-degree angle, and is as tall as "½" the width of the building.

If the plans called for a 5 in 12 roof, you'd know immediately that the roof will rise 5 inches for every 12 inches it travels horizontally. A 5 in 12 roof will slope at a 22½-degree angle.

The most basic item in a roof is called the *common* rafter. See Figure 12-1. When I speak of the *span* of a rafter, I mean the distance from the

outer edge of the exterior wall to the center of the ridge that the rafter butts into. For most roofs, this is half the width of the building that these rafters span.

In other books you may find the term "span" used to mean the width of the room, but this isn't common jargon on most job sites. We use the term span of the rafter interchangeably with the term run of the rafter. The width of the room is simply called the width of the room. I hope this doesn't run completely counter to all you know about roof framing.

A common rafter reaches from the top plate of an exterior wall up to the ridge board of the roof. Two opposing commons are set exactly opposite each other off the ridge and nailed to the ridge on the same layout. The *ridge* board (Figure 12-1) is the board that's set lengthwise at the uppermost point (or ridge) of the roof. All the rafters nail into the ridge. The layout on the ridge board matches the layout on the tops of the exterior walls so that the rafters are straight.

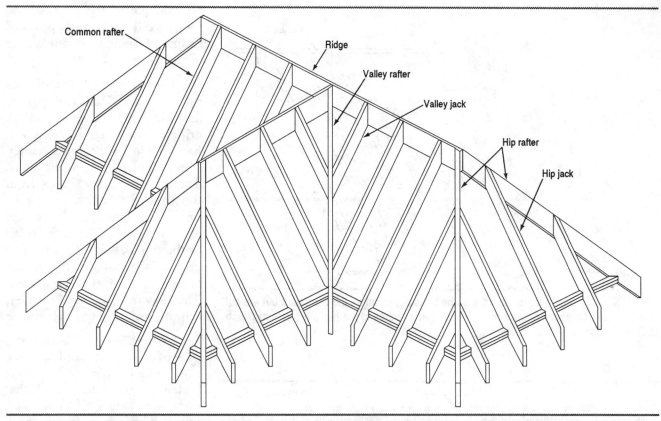

Figure 12-2   Here's a typical hip roof. Notice the relationships between the hips, valleys, common rafters, jack rafters and ridges.

A *plumb cut* (or *top cut*) on a rafter refers to the uppermost cut that's made on the rafter so it fits flat along the ridge. You can easily find the plumb cut of a common rafter using your rafter framing square or a speed square.

A *birdsmouth* is the cut made on the lower end of the rafter where it rests on the exterior wall (Figure 12-1). The birdsmouth has two cuts. There's a short plumb cut that's in line with the outside edge of the exterior wall and parallel with the top plumb cut. The seat cut intersects with the plumb cut. It marks the place where the rafter rests on the wall. The amount of wood that's left above the seat cut (in line with the plumb cut) is called the *throat*.

A *valley* rafter is where two planes of a roof meet and form a valley, usually on an inside corner. It's called a valley because it forms a major point of drainage for the roof. See Figure 12-2. The rafters that extend from the ridge to the valley rafter are called *valley jacks*. The plumb cut on the valley jack is the same as the plumb cut on the common rafters.

A *hip* rafter is also a point of intersection for two perpendicular roof planes (Figure 12-2). It runs diagonally down to the outside corner of two perpendicular exterior walls. The rafter that runs from the wall to the hip rafter is called a *hip jack*.

A hip jack has the same birdsmouth cut as the common rafter but the plumb cut is adjusted to meet the hip. When a jack attaches to a hip or a valley, it does so at a 45-degree angle. But because the intersection takes place above the flat plane of the walls (at different heights depending on the pitch of your roof), the angle cut you make along the plumb cut of the jack at the jack/hip or valley connection will vary depending on the pitch of the roof. This angle that forms your plumb cut is a side or cheek cut. See Figure 12-3. The cheek cuts on hips and valleys are

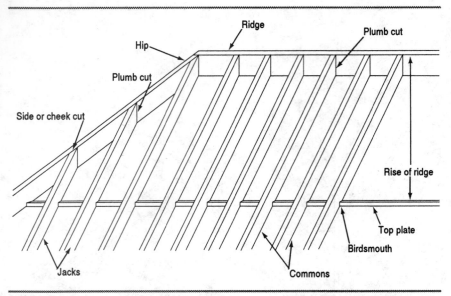

Figure 12-3   Jack rafters have the same birdsmouth cut as the commons but are cut shorter where they attach to the hip. The cut that attaches to the hip is a common plumb cut with an angled side or cheek cut.

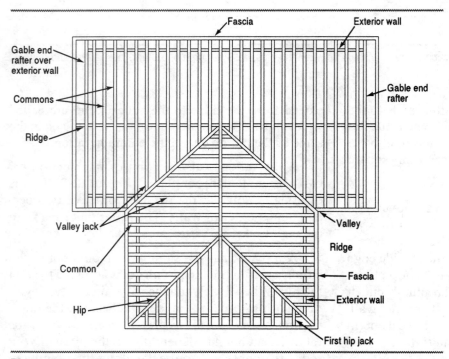

Figure 12-4   Some roofs will have a combination of hip and gable ends. This roof has two gable ends and another section that's designed with two hips. The gable ends have commons that end directly over the exterior wall line. The hip section has two hip rafters set diagonally from the wall corners and intersecting at the ridge. These two perpendicular sections of roof intersect at two valleys.

also unique to the pitch of the roof. We'll go through the procedure of finding this angle later in the chapter.

There's a third type of jack rafter besides the valley jack and hip jack: a cripple jack. A cripple jack occurs where you have a hip rafter and a valley rafter right next to each other. The cripple jack extends from the hip to the valley. This is the only rafter that doesn't connect at some point either to the ridge or the top plate of the wall. The cripple is in essence "flying" between the hip and the valley. That's why you'll often hear cripples referred to as flyers.

There are three basic types of roofs: a *lean-to* roof, a *gable* roof and a *hip* roof. A lean-to is just a single planed roof that goes from one wall to another, with no change in direction. A gable or a hip roof changes directions at ridges, valleys, hips or dormers. A gable roof may use a few hips, as shown in Figure 12-4. And then again, a hip roof might be found with a gable end. Both designs would make liberal use of common rafters, as in Figure 12-4.

Besides the lean-to, a gable roof is the most basic of roof designs. A gable roof uses common rafters throughout the body of the roof and terminates plumb over each end wall. Take a look at Figure 12-5. A room may be built perpendicular to the main body of rafters, which would require a separate group of common rafters perpendicular to the main rafters, but these too would terminate over any end wall with two common rafters. Two valleys and their accompanying jacks would connect the separate bodies of perpendicular rafters.

A hip roof has diagonal hip rafters that extend from the outside corner of two perpendicular exterior walls up to the ridge board (see Figures 12-2 and 12-4). You usually see two hips come from the opposite exterior corners and meet at the ridge. In more complicated designs you might have a hip set all by itself, but usually they come in pairs.

The *total rise* of a roof is the height of the tallest ridge in the roof, *measured from the first floor* (or *grade* as it is commonly referred to). This is very important in communities that have building height limitations. Don't confuse this with the *rise of the ridge*, which is the height of the ridge above the top plates of the walls (Figure 12-3). The rise of the ridge *plus* the height of the walls below it (all the way down to grade if you have two stories) will give you the total rise of the roof.

A *dormer* is a small set of walls with a gable or hip roof that's attached to the side of another roof. Take a look at the gable dormer in Figure 12-6. It also has a window built into it. Like any roof, its height is calculated from the width of its miniature walls. Its main function is to let light into an upstairs room or attic. Dormers are an integral part of the design on certain styles of houses, especially in the Northeast.

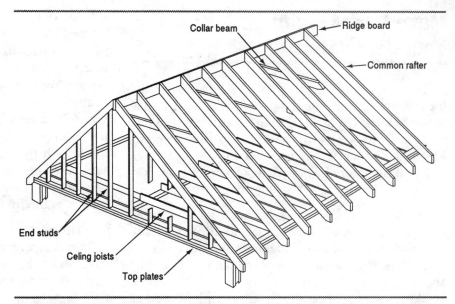

Figure 12-5 A straight gable roof will contain only common rafters with no intersecting valleys or hips. The commons are laid out from one end right through to the other. They end directly over the exterior walls.

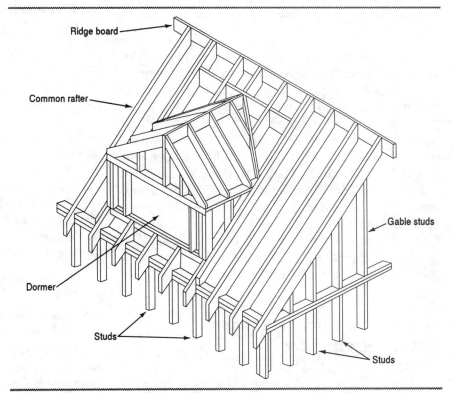

Figure 12-6 Dormers are small walls with roofs built on them that allow light into attics or upstairs bedrooms. Typically they "pop-out" of larger sections of steeply-pitched roofs.

Make Friends with Your Calculator

For working on the job site, I use a small, inexpensive solar calculator to cut all my roofs. When it starts to show signs of wear, I put it in my truck and buy a new one so I always have a spare. I like the regular old compact model with the vinyl cover. It fits in my bags with my other tools, so it's there when I need it, not sitting in my truck.

I use the cheap ones because all I've ever needed to cut a roof is the simple addition, subtraction and multiplication functions they offer. I've studied the trigonometry involved with roof cutting and find that for 95 percent of the work I do, I don't need it. The only specialty function I might need is the square root function. When I do need these more exotic functions, the work is so critical that I'm working at a snail's pace anyhow. If I had a really trick calculator, I'd be too afraid to take it 20 feet up and possibly drop it. So it'd probably spend all day in my truck keeping my rafter length book company.

There are new calculators on the market that do anything from figure concrete to calculate how much smaller your check is going to be each week because you own it. They're user-friendly, since you can enter your numbers in feet and inches without converting to decimals. But I think they're just expensive toys. I'll show you how to convert feet to decimals easily. Then if you just *have to have* one of those fancy gadgets, go for it. But in my opinion you're better off putting the cash into sensible tools that are *really* going to help you earn a living. If you don't have your eye on any tools you need, I'll suggest a couple later in the chapter, so hang onto your cash for a minute.

■ Converting from Fractions to Decimals

One of the major stumbling blocks for framers learning to use a calculator is converting measurements from feet, inches, and fractions of an inch into a decimal that the machine can use. It's really quite simple. The first trick is to always read your original measurement in *inches only*, and not feet and inches. You may convert back to feet later, but always read your original number in inches. For example, instead of reading the tape as 6'6½", train yourself to read 78½ inches. Then enter 78.5 into the calculator. Of course, you could enter 6 feet, multiply it by 12, then add 6.5 inches to reach the same conclusion — but why bother? It's the long way home. Teach yourself to enter your numbers in inches as the first step.

The next step is to memorize the conversions of fractions of an inch into decimals. If this causes you to flash back on the math classes you hated, I'm sorry. If the trauma was so great that your brain just can't absorb these innocent decimals, then do what I did. Tape a little cheat sheet to the back of your calculator. Sooner or later, without your being aware of it, you'll discover that you've learned them.

Here are the most common conversions you need for rough framing:

$$1/8 = .125$$
$$1/4 = .25$$
$$3/8 = .375$$
$$1/2 = .5$$
$$5/8 = .625$$
$$3/4 = .75$$
$$7/8 = .875$$

If you have to find the decimal equivalent for a fraction, just divide the top number of the fraction by the bottom number to get the decimal. Try it with ⅞. If you divide 7 by 8, you'll get 0.875.

When we get around to figuring rafter lengths, you'll want to be proficient at converting fractions into decimals. For instance, you'll constantly be dividing the width of the room in half to find the distance a rafter actually spans. Rarely will you have a room that has exact dimensions. If you aren't quick with the decimals, or worse, if you enter the wrong decimal, you'll waste a lot of time as well as lumber. Once you're proficient with your calculator, you can easily enter the width of any building, no matter if it's a whole number or a fraction. If a room measures 25′5¾″ wide, enter it in your calculator as 305.75 and divide by 2 to find the span of the rafter: 152.875 inches.

Another reason you'd better get used to working with decimals is that within the next few years we'll be required to use metric. Even the building code people are starting to give measurements in millimeters. We'll soon be spacing studs 400mm on center. Decimals and the metric system go hand in hand.

Using the Framing Square

The next tool you'll need if you plan on cutting many roofs is a framing square. There is no other tool that holds so many solutions to roof layout problems. Many framers are afraid of the framing square. They seem to feel that you need a PhD in mathematics to figure out all those little numbers.

I understand that it's useless for me to hype the square as the cure for all roof framing problems — and then leave you to figure it out yourself. If you're anything like me, you'll never get around to it. But I've figured out all those little numbers and I'll set you straight on the ones you need (and even let you in on some secrets held by the ones you'll never use).

For roof cutting, you need a square with tables that give the length of common, jack and hip rafters, per foot of run. I use the *Stanley Model 45-020*. It has everything you need to figure all the cuts on a roof, including an instruction booklet that explains all the functions of the square.

Some of the functions on the square are either outdated or seldom used by the average carpenter in this age of calculators and computers. But there are many useful and easy-to-understand tables that I'll explain. Without these tables the square won't help you, and neither will this chapter. So do yourself a favor and spend the money you saved on that fancy calculator on a good square. Believe me, it's worth it.

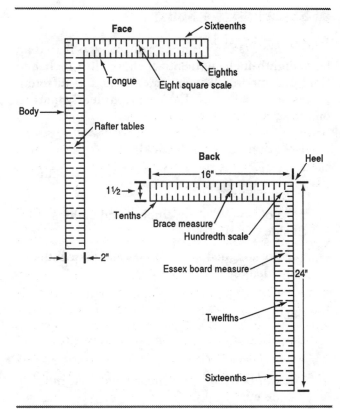

Figure 12-7 The Stanley rafter framing square is probably the most valuable tool you could own for figuring the cuts on a roof. It's divided into the *face* and the *back* on the 2-inch-wide by 24-inch-long *body*, and the 1½-inch-wide by 16-inch-long *tongue*. The body and tongue intersect at the *heel*. In strategic locations throughout the square you'll find an assortment of tables and scales that are invaluable for calculating a roof.

The framing square is divided into two main parts: the body and the tongue. The body is the 24-inch long, 2-inch wide blade on the square. It usually represents the level, horizontal line when laying out rafters. The tongue is 16 inches long and 1½ inches wide. It represents the rise or plumb line when laying out the rafters. The body and the tongue of the square intersect at the heel. See Figure 12-7.

When you hold the square with the body horizontal, the heel in your right hand and the tongue extending *up*, you're looking at the *back* of the square. If you hold the body horizontal, with the heel in your right hand and the tongue extending *down*, you're looking at the *face*.

Scales on the Square

On the Stanley framing square, you'll find an inch, eighth-inch, sixteenth-inch, twelfth-inch, and finally a hundredth-inch scale located in different places around the tool. Each scale is clearly marked on the square:

Location	Scale
Face of body (outside edge)	Inches and sixteenths
Face of body (inside edge)	Inches and eighths
Face of tongue (outside edge)	Inches and sixteenths
Face of tongue (inside edge)	Inches and eighths
Back of body (outside edge)	Inches and twelfths
Back of body (inside edge)	Inches and sixteenths
Back of tongue (outside edge)	Inches and twelfths
Back of tongue (inside edge)	Inches and tenths

Whenever possible, use the outside edge of the square on both the body and the tongue. This keeps confusion to a minimum and makes it easier to handle and balance the square. For all my examples, I'll always assume you're using the outside edge of *both* the body and the tongue.

If you're laying out a cut and you mistakenly use the outside edge of the tongue and the inside edge of the body, you'll make an inaccurate mark, which will lead to an inaccurate cut. If you need to use the inside edge for scale purposes, use the inside edges of *both* the body and the tongue.

These scales are helpful for stepping off rafters, marking cheek cuts, marking blocks, and for general measuring purposes. Although most rough framers don't bother with sixteenth scales (except for exposed cuts), the eighth and the twelfth scales are extremely helpful. The only problem with the twelfth scale is that it doesn't show eighth markings, which, for instance, makes it difficult to mark out 10⅜. No problem! Just flip the square over and you'll have the sixteenth scale on the outside edge. There you'll find all the eighth markings you'll ever need.

The hundredth scale has been largely replaced by the calculator, and on some squares it has been omitted. If you do have it on your square, it can be a lifesaver if your calculator prematurely dies. You can use it to convert fractions into decimals. This scale is found at the heel on the back of some squares. It's a short scale with 100 small lines on top, and a sixteenth scale with (you guessed it) 16 lines on the bottom. The hundredth scale is also broken up into four divisions of 25, which make it easier to read. The three lines that mark these divisions line up with the three lines on the sixteenth scale that represent ¼, ½, and ¾ inch.

Here's how to use the hundredths scale to convert a fraction to a decimal. Let's start with an easy one, say ¾. First you have to convert it to sixteenths by multiplying both the top and the bottom number by 4: ¾ converts to ¹²/₁₆. Now, starting on the right side of the sixteenths scale, count off ¹²/₁₆. You'll be lined up with the third large line on the hundredths scale. Since the large lines each equal 25, and you're on the third one, this equals 75/100. In decimal form, 75/100 is 0.75. So the fraction ¾ equals the decimal 0.75.

Tables on the Square

One table you may find useful on the square is the *Brace Measurement Table,* which is the group of numbers down the center on the back of the tongue. This is a quick guide for finding the lengths of braces (or short rafters for that matter) that hold post or wall members plumb and square with the ground or the surface they are rising from. This table works on the familiar right triangle formula: $A^2 + B^2 = C^2$.

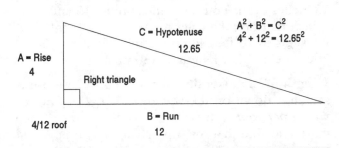

Figure 12-8   The theories behind roof cutting revolve around the properties of the right triangle. For a 4 in 12 pitch roof, the rafter rises 4 inches for every 12 inches it travels horizontally. Applying the $A^2 + B^2 = C^2$ formula, you'll find that the hypotenuse is 12.65. So a rafter set on a 4 in 12 pitch will be 12.65 inches long for every 12 inches it spans.

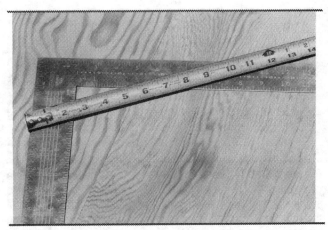

Figure 12-9   You can use a tape measure and your framing square to double-check the exercise in Figure 12-8. Hold the end of your tape on the 4-inch mark on the body and then line it up with the 12-inch mark on the tongue. You'll notice that 12⅝ (12.65) reads on the tape when it lines up with the 12-inch mark on the square.

If you look closely at this table you'll find the first set of numbers on the left-hand side are the numbers 24 stacked on top of each other, followed closely by the number 33.94. This means that for a right triangle with the bottom and the side at 24 inches, the hypotenuse is 33.94 (24^2 x 24^2 = 33.94^2).

If you were supporting a post with a brace, you could use this scale to find that if you set it 24 inches up the post and 24 inches away from the post, the brace would have to be 33.94 inches long for the post to be plumb and square with the ground. This works because of the laws of a right triangle.

These numbers also work for a rafter because rafters work on the same principle. If you set a rafter 24 inches high and 24 inches away from the building, it will be 33.94 inches long and set at a 45-degree angle. Which leads us back to where we began this discussion — cutting a roof!

■ Rafter Tables on the Square

When using a framing square for roof layout, we rely on the same formula that we used back in Chapter 4 when we squared up the walls as we snapped them on the floor. Let's say that the plans call for a 4 in 12 roof. The plans are telling us that

for every 12 inches the roof travels horizontally, it rises 4 inches vertically. So what would the hypotenuse of that triangle be? That's an easy one:

$$4^2 + 12^2 = C^2$$

$$16 + 144 = 160$$

Then find the square root of 160 to find the hypotenuse of 12.65. This is shown in Figure 12-8.

What this means is that the roof travels on an angle 12.65 inches for every 12 inches it travels horizontally. To prove this to yourself, get your framing square and follow me through the exercise that's illustrated in Figure 12-9. Hold the end of your measuring tape at 12 inches on the tongue and form a hypotenuse by angling up to the 4-inch mark on the body. Notice how the 12⅝-inch mark (12.65) on your measuring tape lines up with the 12-inch mark on the tongue.

If you followed me through this simple exercise, congratulations, you've just learned all the math you'll need to calculate the cuts for a roof. Everything that follows will expand off this simple formula.

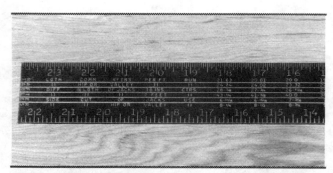

Figure 12-10   The table you'll use most often on the square is the rafter table on the face side of the body. It's divided up into six categories. Use the inch markings along the outside edge of the square for the roof pitch (4 inches for 4 in 12, 5 inches for 5 in 12, etc.) and use the numbers listed under them. Simply line up the number with the sentence you're working with on the square.

Length of common rafter per foot run. Hold the square with the body horizontal, with the tongue dropping off to the right. On the far left corner of the square you'll notice six sentences. The first sentence should read *Length of common rafter per foot run*. To the right of this sentence, you'll see a list of numbers. Look at Figure 12-10. If you look closely, you'll notice the numbers line up underneath the inch numbers on the top outside edge of the square.

For the first sentence, these smaller numbers represent the length of a common rafter, spanning 1 foot, at the pitch indicated by the inch number on the outside edge it's lined up under. If you look under the 4-inch mark on your square, you'll see that a rafter is 12.65 inches long for every foot it spans.

Well, what do you know? That's the number we calculated! It's right there on your square whenever you need it. And now you know how to calculate that number if your square ever disappears.

For a 5 in 12 pitch, the length per foot of run is 13 inches even. You'll see the number 13 under the 5-inch mark on the square. To find the length that it travels in 10 feet, simply multiply 13 by 10 and you'll get a rafter length of 130 inches.

Length of hip or valley per foot run. The next sentence down on the square is *Length of hip or valley per foot run*. Again, look under the 4-inch mark and find that the hip or valley will be 17.44 inches long for a 1-foot span on a 4 in 12. Hips and valleys for a particular span will always be the same length. The distance from the top plate to the main ridge is exactly the same for a valley or hip.

You'll notice that a valley and hip are longer per foot of span than a common rafter. That's because they leave from a point that's farther away from the ridge than a common. They're not only longer, but they're also set at less of an angle so they plane in with the commons.

The angle and position of a hip in relation to a common rafter is clearly shown in Figure 12-11. The line from A to B represents the length of top

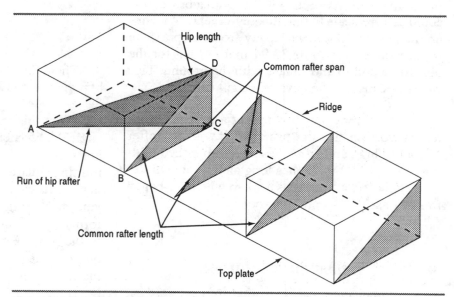

Figure 12-11   The relationship of the hip to the common rafter can be clearly seen inside this three-dimensional cube. Line AC is the run of the hip rafter, AD is the hip itself, BC is the common rafter span, and BD is the common itself. Notice that the hip line AD is longer than the common line BD. But they end up on the same 4 in 12 (19 degree) plane. This is because a hip travels 17 inches for every 12 inches a common travels, regardless of roof pitch.

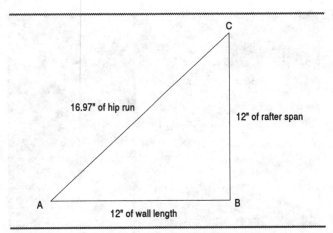

Figure 12-12 This right triangle demonstrates how we came up with the 17-inch run for the hip. The lines AB, BC, and AC were all taken from Figure 12-11. AB is the wall line, BC is the common rafter span, and AC is the horizontal run of the hip. Using the A^2 + B^2 = C^2 formula, we find that the hypotenuse is 16.97, commonly rounded to 17 inches. So now we know that a hip travels 17 inches horizontally for every 12 inches that a common travels horizontally, regardless of the pitch.

Figure 12-13 You'll always mark out a hip plumb cut using the 17-inch mark instead of the 12-inch mark on the square. For a 4 in 12 roof, hold the 4-inch mark on the outside edge of the tongue and the 17-inch mark on the outside edge of the body, and then strike a line along the tongue. In this photo the square is on the lower edge of the hip in order to get the plumb cut angled properly with the crown. It makes no difference which edge of the material (top or bottom) you use.

plate from the corner of the building to the position of the common king rafter, which is placed at the point where the hip meets the ridge. The line from B to C is the span of the common rafter. The line from A to C is the run of the hip. If we take these three lines as a triangle, with AB and BC as two sides and AC as the hypotenuse, we could determine the "length of hip, per foot run."

If AB and BC both measure 12 inches, what's the length of AC? Take a look at Figure 12-12. Going through the math, we end up with $12^2 + 12^2 = 16.97^2$. This is usually rounded up to 17. So for every 12 inches that a common rafter travels horizontally towards the ridge, a hip rafter travels 17 inches. You'll calculate the hip on a 4 in 12 roof at 4 in 17.

So when laying out a hip or valley, you can use the 17-inch mark on the body of the square instead of the 12-inch mark. When making the top or bottom plumb cut mark on a 4 in 12 roof, hold the square at the 4- and 17-inch markings along the outside edge of the tongue and body and scribe along the tongue to get the mark. Check out Figure 12-13. A 5 in 12 hip or valley is made on a 5 in 17, and so on.

On a speed square, there are different areas to read for a common rafter and a hip/valley rafter, stamped right into the steel.

Difference in length of jacks at 16-inch and 24-inch centers. The next two sentences are *Difference in length of jacks at 16-inch centers* and *Difference in length of jacks at 24-inch centers*. These are the lengths of the first jack rafter as well as the amount you would add to each successive jack rafter to keep them on layout moving up a hip or valley.

Let's say you have a 4 in 12 roof laid out at 16 inches on center. According to the number under the 4-inch mark on the square, your first jack rafter would be 16⅞ inches long. The next jack would be twice as long, or 33.75 inches. The third jack would be three times as long, or 50.625 inches. You can multiply as I just did, or you can simply add 16.875 inches to each jack to get the next one.

It's easy to see what a timesaver this method is. You can make every one of your jack rafter packages the same. If your common rafters are a couple of feet longer in one section, you'll just have one

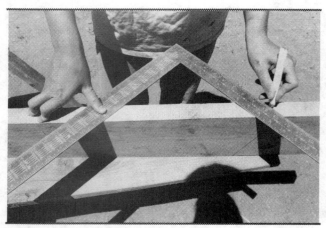

Figure 12-14   Marking out a side cut is easy with the framing square. After the plumb line is made along the face, use the top edge of the rafter to find the side cut angle off the square. The rafter table on the square will give you the number to hold along the tongue (11⅜ for a 4 in 12 common), and you'll always hold 12 along the body. Line up the side cut with the plumb cut, as the side cut will actually be made by the saw blade set at an angle.

Figure 12-15   Once the side cut is marked out with the framing square, you can find its exact angle with the speed square. You don't need to make any more side cut marks. Just set your saw to this angle each time you make the plumb cut.

more jack here than elsewhere. Even though a jack is attached to a hip or valley, it still gets a common (4 in 12) plumb cut on it and not the hip or valley plumb cut (4 in 17).

Side cut of jacks.   The fifth sentence down is *Side cut of jacks use*. When a jack rafter hits a hip or valley, an angled side or cheek cut must be made along the plumb cut to attach it cleanly to the hip or valley. (Look back to Figure 12-3.) This line gives the angle of this side cut. For a 4 in 12 roof, look under the 4-inch mark and you'll see the number 11⅜ on the fifth line. But how does this relate to an angle cut?

The angle you'll cut is marked out along the top edge of the jack rafter. This is also the angle that your saw blade is set when cutting the plumb cut. When you use the framing square to make this mark, you'll hold the two outside edge inch markings on the tongue and the body along the top edge of the rafter; the number 12 on the body and the number 11⅜ on the tongue, as shown in Figure 12-14.

Keep in mind that the square may have to be flipped in order to make the side cut match the direction you need for your particular hip or valley.

Also, you'll probably already have your plumb cut marked out for the long point of your jack. When making your side cut mark on the top edge, you'll need to slide the square down the top edge of the rafter until the tongue intersects the top point of your plumb cut. Make sure you have 12 and 11⅜ lined up on the rafter edge, then make your mark.

After you've marked out one jack, you can use your speed square to find the exact angle that you just made. All the jacks will be cut with the saw set at this angle. For example, the mark that we just made using 12 on the body and 11⅜ on the tongue would produce a 43-degree angle. See Figure 12-15. Once you know this angle from the speed square, there's no need to use the framing square to find it again.

Side cut hip or valley.   To find the side cut for a hip or valley, use the sixth line which reads *Side cut hip or valley use*. When a hip or valley approaches a ridge, it does so at an angle. Use this line to find the angle of the plumb cut that will attach the hip or valley to the ridge. Because it's a hip or valley, the side cut is different from a common jack. Make these side cut marks exactly like the jack cuts (as in Figures 12-14 and 12-15), except use the numbers from the sixth line on the framing square.

For example, on the 4 in 12 roof, look at the column under the 4-inch mark on the outside edge of the square, and read the number on the sixth line — 11¹¹⁄₁₆. Hold the square along the top edge of the hip/valley material and line up the number 12 on the body and 11¹¹⁄₁₆ on the tongue. After making sure you're also lined up with the long point of the plumb cut, make a mark along the tongue on the top edge of the hip/valley. After you make the mark, use your speed square to find its exact angle. Using the numbers 12 and 11¹¹⁄₁₆, I come up with a 44-degree angle.

That does it for all those little numbers on the framing square. Now we can take a few more steps towards that lumber pile. But leave that saw in your truck! We're not ready to ruin any lumber yet. Let's get the numbers down first.

Calculating the Roof Package

When you're cutting a roof, you always begin by finding the actual width of the building framing, regardless of what the plan says it should be. Jump up on the top plates and take the exact measurement from the outside of one top plate to the outside of its opposite. Take the actual width of the building or room that this section of rafters will stack over and divide it in half to find the run or span of the rafter.

If two rafters oppose each other off a common ridge that's centered between them, and they're both sitting on walls of the same height, their spans as well as their calculated lengths will be the same. So your calculations will work for both sides. You'll also use this span to figure any hips or valleys.

In the calculations that follow, we'll assume that the width is 25′5¾″ (or 305.75 inches), which, divided in half, would make the rafter span 152.875 inches. The roof will be a 4 in 12.

■ Calculating the Common Rafters

Always begin your calculations with a common rafter. The length of a common rafter per foot run for a 4 in 12 roof, according to the square, is 12.65

inches. For every "12 inches of span" on a level horizontal plane, the rafter is 12.65 inches long on a 19-degree angle. Of course "12 inches of span" is 1 foot. So if we know how many feet (or fractions thereof) there are, we can multiply that by 12.65 to find the rafter length.

How many feet are there in the span of our sample building? We know there are 152.875 inches. To convert inches to feet, divide by 12: 152.875 ÷ 12 = 12.739. So the span from the outside edge of the wall to the center of the ridge is 12.739 *feet*.

To find the rafter length, multiply 12.65 inches by 12.739 feet to arrive at 161.155 *inches* (12.65 x 12.739 = 161.155). I know it sounds crazy to multiply inches by feet, but that's not really what we're doing. We're multiplying one number in inches by another number that just happens to be expressed in feet. If it makes it easier, consider the 12.739 to be *units* instead of *feet*.

We've calculated the common rafter to land perfectly in the center of the room. But it actually meets the ridge board, which is in the center of the room. So you'll have to subtract half the thickness of the ridge from the calculated rafter length. If the ridge is 1½ inches thick, subtract ¾ inch (0.75) from 161.155 inches. That makes the common rafter length 160.405 inches.

The entire sequence on the calculator would go like this:

$$305.75 \div 2 = 152.875$$

$$152.875 \div 12 = 12.739$$

$$12.739 \times 12.65 = 161.155$$

$$161.155 - .75 = 160.405$$

This takes me 15 seconds on my calculator, and about 15 minutes from a rafter table.

As rough carpenters, we would mentally round this off to 160.375, as we're more fond of working with eighths than sixteenths. And take a tip from me: Write out your calculations on the first common rafter you cut. When you're up on the walls setting ridges and something isn't working, it's nice to have everything spelled out. You can see if you calculated it wrong when you cut it. I've developed a shorthand method of writing down my calculations:

305.75

152.875

12.74

161.155

160.375

After I've double-checked the numbers, I'll write them in this form on the first common. I don't find any reason to include my division or multiplication symbols or the number 12.65 because I know the sequence only too well. This is quicker and easier than writing out the whole calculation, and just as effective.

■ Calculating Hip and Valley Rafters

To calculate a hip or valley, you use the same procedure, except you'll use the number on the square under *Length of hip or valley per foot run*. For the sample 4 in 12 roof, that's 17.44 inches. The span is the same: 12.74 feet.

So the calculation is: 12.74 x 17.44 = 222.185

The hip/valley length is 222.185 inches. Now we need to subtract for half the thickness of the ridge. Because a hip or valley approaches the ridge at a 45-degree angle (Figure 12-2), subtract the distance diagonally, along this angle, to the center of the ridge. Take your speed square and draw a 45-degree angle along the top of the ridge board. Now mark out the exact center of the ridge on this 45-degree mark. If you measure along this line from the edge to the center you'll find the amount to subtract from your hip. For a 1½-inch-thick ridge, this is about 1 inch. For a 3½-inch-thick ridge, it's about 2⅜ inches. This number remains the same regardless of the pitch.

For our example, shorten the hip/valley length by 1 inch, and you end up at 221.185 inches. Round this off to the nearest eighth, making it 221.125. Again, I suggest writing down your calculations on the actual hip/valley for later reference.

■ Calculating Jack Rafters

Calculating a hip or valley jack package is easy once you know the length of the first jack. But how do you measure the first jack if the hip isn't set? Remember, the framing square gives the length of the first jack. And you add this number to the previous jack to get the next jack in line. For the 4 in 12 roof, the square tells you that the first jack is 16⅞ inches long. Again, subtract half the thickness of the hip or valley along the diagonal, which is 1 inch for 2-by material. So the first jack is 15⅞ inches long.

But there's one more thing to keep in mind here. When a jack is set on a wall and travels up to a hip or valley, it creates a triangle. Look at the triangle in Figure 12-16. To figure the length of the jack to the center of the hip, subtract half the diagonal thickness of the hip to find the exact length of the jack.

But remember I said that the triangle goes down the center of the top edge of the jack. So when we say that the jack is 15⅞ inches long, we mean that the *centerline* of the jack is 15⅞ inches long. This is important because the jack has an angled side or cheek cut. The jack should be 15⅞ inches to the middle of the angle. But this is a confusing way to tell your buddy what length jack to cut. Since the jack is 1½ inches thick, add half the thickness of the stock (¾ inch) to get a long point, or subtract it to get a short point. It's always easier to cut an angle from the short point, so let's subtract ¾ inch and arrive at a jack length of 15⅛ inches to the short point of the plumb cut.

As long as we're on the subject, let's look at the other line of our triangle: from the hip to the jack. This is important because it will give us the point where the first jack goes on the wall. Remember that the square originally gave us a jack length of 16⅞ inches. If our square is a true right triangle, this same 16⅞ would be the distance from the corner the hip is resting on to the middle of the jack. But you want to lay out marks for both edges of the jack, not the middle. So subtract ¾ inch to make a mark at 16⅛, then add ¾ inch and make a mark at 17⅝. These marks show exactly where the jack rafter goes. Figure 12-16 shows this clearly.

The only time you may run into a problem is if the ceiling joists are laid out without considering the jacks, which is very likely. At this point you have to work backwards from the actual measurements. Let's say that the ceiling joist is laid out between 16 and 17½ inches from the corner. To lap along the side of the joist (which is imperative) the jack would lay out on the wall between 14½ and 16 inches. So what's the jack length?

First find the center of the layout, which in this case is 15¼ inches. So if one side of the triangle is 15¼ inches, we need the other side (the jack length) to be 15¼. So the jack is 15¼ inches long, minus the diagonal on the ridge (1 inch for 2-by stock), or 14¼ down the middle of the jack. Subtract ¾ and you get 13½ inches to the short point.

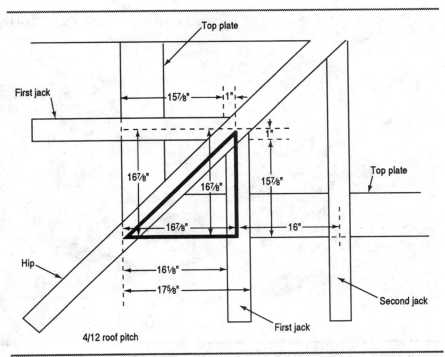

Figure 12-16   When a jack is set to a hip or valley, it creates a triangle. By studying the relationship of the parts of the triangle, you can determine the layout of the jack and its length in case you need to make a special one. Subtract half the diagonal thickness of the hip from the first jack set. After that you can add the "difference in length of jacks" found on the framing square (16⅞ inches for a 4 in 12 at 16 inches on center) to this adjusted number and come out with a properly-cut second jack.

Now to find the lengths of the remaining jacks, check your square on the line reading *Difference in length of jacks at 16-inch centers*. Look under the number 4 to find 16⅞ inches. To find the length of the next jack on layout, add 16.875 to 15.125, and you'll get a second jack at 32 inches, also to the short point.

Remember, you subtracted the diagonal on the hip on the first jack so you won't need to do it again. Just keep adding 16.875 and make a cut list: 15.125, 32, 48.875, 65.75, 82.625, 99.5, 116.375, 133.25, 150.125. The first jack was 15.125 sp (short point) and the last jack is 150.125 sp, for a total of nine jacks to cover one side of the hip or valley. The other side would have the same jacks, but with the short point on the opposite face of the jack.

You calculate valley jacks just like hip jacks, except you can start at any number that allows somewhere around a 16-inch final bay. It's virtually impossible to land a valley jack on layout with the incoming commons unless you measure its length after the commons are up, which is how some people do it. Why? Because they're concerned about keeping a perfect 16- or 24-inch on center layout for the entire roof. But is this necessary?

In my experience, no. The only person who will be affected by this decision is the sheather. When an experienced sheather approaches a valley, he's usually laying down full sheets, and when he gets to the point where a full sheet won't fit any longer, he stops. He'll then continue in the same manner with the next row. The next day, after all the full sheets are down, he'll take out his saw and do all the custom cuts, which inevitably involves each and every valley. The very experienced sheathers are used to being slowed by valleys and they have tricks (like lapping sheets) that fly right over sections where a layout changes.

Figure 12-17   If you have quite a few commons to cut, place them crown down in a rack like the one pictured. This building had no fascia, so the rafters didn't need a tail. The end that's pictured has the seat and the plumb wall line cut. The seat was cut with a swing table on a circular saw, and the wall line cut was made with a chain saw jig on a circular saw.

I suggest using the same progression as the hip jacks to make it easier on yourself and the stackers. If you keep the same progression for cutting all of your jacks, you'll greatly reduce the chances of error — both for you and the stackers.

There's one exception to this rule. If you have exposed rafters, you definitely want the layout to be as near to perfect as possible. You should wait to cut the jacks until the commons and valleys are up and you can take a precision measurement.

Cutting the Roof Package

When calculating and cutting the roof package, you might find it helps to calculate each separate span right before you cut it. Some folks like to sit down at night to do all the calculating at once, and show up the next day with a list to cut. I've always preferred doing my math right on the material I'm

working with. Then I can either cut it myself or direct my laborer. If, for some reason, I do some calculating ahead of time, I always double-check and transfer my math onto the material. Some might consider that overkill. I don't.

When preparing to cut, you can save time by setting up racks to support the groups of common rafters if you have enough to make it worth it. Check out the easily-built rack in Figure 12-17. Don't spend all day making it. It doesn't need to be anything fancy. Just take two pieces of rafter stock and nail some blocks to their ends so they won't roll over on your toes, and then toenail some scrap blocks on top to support the group of rafters. Once all the rafters are on the stands, make sure they fit tightly together, then add another block on top to keep them standing. Always spread the rafters in the rack upside down (crown down) as you'll need access on the underside to make the seat cut.

Stop here — before you rack up all your rafters. Don't do it unless you have a swing table for your saw. A swing table is a table that allows your blade

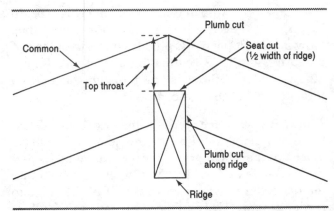

Figure 12-18   For rafters designed to sit on top of the ridge, you'll need to make a plumb cut down to a seat cut, and then another plumb cut below it. The top plumb cut is determined by the calculated throat. The seat is half the thickness of the ridge to allow its opposing rafter to rest on the ridge as well. The lower plumb cut ends up resting along the ridge.

to swing to angles beyond 45 degrees, the maximum for a normal table. The seat is cut at the complementary angle of the pitch of the roof. A 4 in 12 is 19 degrees, so the complementary angle is 71 degrees (90 - 19 = 71). The seat cut is cut at 71 degrees, which you can't cut with the normal table on your saw. If you don't have a swing table, don't waste your time racking up the rafters, I'll show you how to cut them individually.

For cutting on a rack, you'll also need either a "suicide blade" or a chain saw jig for your circular saw to make the top cut. You can find them at any well-stocked contractor's hardware store. The suicide blade is a setup that mounts the blade outside of the guard, so you can make the top cut without having to move each rafter in the rack. The chain saw jig basically mounts a bar with a chain on it to your circular saw. The bar adjusts so you can set the proper top cut angle. Expect an investment of around $200 for the swing table and suicide blade and closer to $450 if you add the chain saw jig.

It's not unusual for a very cut-up roof to have few commons and a whole lot of jacks. When that's the case, you might be better off making a pattern and tracing each individual rafter. Setting up a couple of sawhorses and placing a man with a saw on either end of the rafter works well.

Figure 12-19   A common plumb cut is easy with a framing square. Simply hold the square on the face of the material. Depending on the crown of the wood and the direction of the plumb cut, you can work off the lower or the upper edge of the material. Use the inch markings along the outer edge of the square. Lining up the numbers with the edge of the material, hold the body on the 12-inch mark and the tongue on the 4-inch mark. Scribe a line along the tongue and there you have it.

■ Cutting the Common Rafters

After double-checking your calculations, go ahead and start cutting the commons. The top cut on a common is made one of two ways, depending on the design of the roof. A common can meet a ridge with the top of the rafter matching the top of the ridge (Figure 12-1). Or it will sit on top of and be supported by the ridge (Figure 12-18). Let's figure the design where the common matches the top of the ridge first.

The plumb cut. With your framing square on the 4 and 12 marks along the outside edges of the tongue and body, make the plumb mark with the long point on the crown edge of the rafter. See Figure 12-19. Or you can use your speed square to make the common 4/12 plumb mark as shown in Figure 12-20. Use the number 4 on the "common column" of the speed square. Put the long point on the crown edge of the board to ensure that the rafter's crowned up when it's rolled. You're now ready to cut.

Figure 12-20   For thin stock (2 x 6), a speed square works great for laying out a plumb cut. Simply hold it on the number 4 in the "common" column for a 4 in 12 common plumb cut. Then just look down at the degree marks to see that this would be 19 degrees.

For a top plumb cut that rests on a ridge, begin by drawing the same 4 in 12 common plumb line. Now measure down this plumb line from the top of the rafter, and mark out the throat size. See Figure 12-18. The top throat should match the bottom throat to make calculations easier.

The seat cut.   Once you have the throat size, measure down the plumb line and make a mark. This is where your seat cut goes. The seat cut depth is half the thickness of the ridge. You're going to butt into the opposing rafter, so leave it half of the ridge to rest on (Figure 12-18). Once you have the depth of the seat cut marked, make another plumb mark off the back side of the seat cut at this point. This second plumb mark will lay along the face of the ridge when the rafter is up.

Once you've marked out the top, you can mark out the rafter for length. Hook on the long point and measure down the top (crowned) edge of the board (Figure 12-21). Make a mark at 160⅜ inches, the calculated common rafter length. Then make another

Figure 12-21   A common is typically measured along its top edge. Hook the plumb cut on top and pull down along the top edge until you have your length. Then make a chicken-foot mark on the top edge.

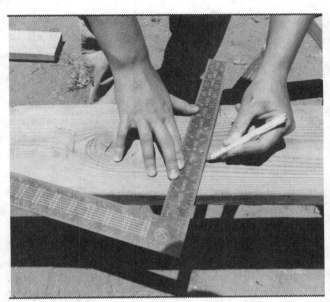

Figure 12-22 Once you have your rafter marked for length, make a plumb mark off the chicken foot. Make this mark, which should run parallel with the top plumb mark (angled the same way), along the entire face of the rafter. This represents your exterior wall line, which is also the plumb cut line on the birdsmouth.

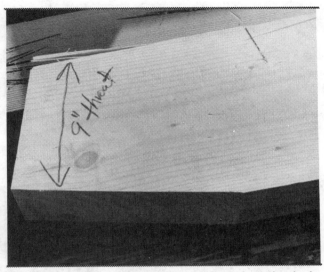

Figure 12-23 The lower stand or *throat* of a rafter is the amount of material that stands above the wall when the rafter is in place. The throat rises off the seat cut and is in line with the outside of the wall when it's in place. This would be a plumb cut.

4/12 mark with 160⅜ inches at the short point, to establish the outside wall line, or plumb cut on the birdsmouth (Figure 12-22). This 4/12 line should run parallel with the top cut.

After you've drawn the plumb line for the birdsmouth, draw in the seat cut. The exact location that the seat cut intersects with the plumb line will be determined by the height of the throat, or stand. The throat is the amount of wood on the rafter rising above the seat cut, and measured on the plumb line of the birdsmouth. Look back to Figure 12-1.

The top throat (if there is one) should match the bottom throat and will be determined one of three ways:

1. If your roof was designed with flush blocks between the rafters on the wall line, make a seat cut to use a standard size block. You don't want to have to rip down every block.

2. The rafter in Figure 12-23 was designed with no tail. The overhang was provided by false 6 x 8 tails applied after the roof was stacked. The tails (measured plumb) were 8 inches tall, and they had 2 x 6 starter board on them. The tails sat directly on the wall with no seat cut. That totals 9½ inches. So we needed to match that with the standard rafters. The rafters had ½-inch plywood roof sheathing on top of them. By cutting the throat at 9 inches, I matched the plywood on top of the rafters with the starter board on top of the false tails so that the roof planed evenly, even though we had a drastic change in material sizes. In the end both framed out at 9½ inches above the top plate.

3. The last possibility is if you have a typical frieze block and fascia setup. At first glance it seems you could set your throat at just about any height. But let me take you a step into the future. If you plan on cutting your rafters in a package, you'll be using a swing table on the seat cut. From my experience, when you use the swing table, the seat cut will have a width of around 2 inches, measured along the bottom edge of the rafter. So if you lay out this 2 inches, your throat will fall where it may.

Figure 12-24   The framing square works well for laying out the seat cut. Use the numbers along the outside edge of the tongue to represent your calculated throat height. Hold the tongue exactly along the plumb line we made in Figure 12-22, and then scribe along the body to mark out the seat cut.

Figure 12-25   Instead of marking all the cuts shown in Figures 12-19 through 12-24 for every rafter, most roof cutters build themselves a jig like the one shown here. You only have to go through the motions once and then you have a pattern to work from. The top plumb cut and the birdsmouth are made exactly the same way. A reference line is drawn where the birdsmouth plumb line intersects with the top edge at the point where you measure a rafter for length —- along the top edge. Be sure to apply a 2 x 4 block centered along the top edge. This helps to keep the top of the jig in line with the material you're working with. Don't cover your reference line with it though!

Now you've drawn the plumb cut line for the birdsmouth. Measure down along this line, from the top edge of the rafter, the distance of the throat. The throat in Figure 12-23 was 9 inches, so you measure down 9 inches from the top of the rafter along the 4/12 line, and make a square mark with the framing square, as shown in Figure 12-24. This represents the point where the seat cut intersects with the plumb line. Then draw a line square off the plumb line and that's the seat cut.

With a rafter that's cut plumb along the exterior wall, you really have no birdsmouth. Since this rafter didn't need a tail, we cut the seat and the entire wall line above it. That just about takes away the birdsmouth. If you have a fascia board or need a tail for some other reason, make sure you cut out only the birdsmouth itself. When making the cut, only cut out the lower part of the plumb line (from the bottom, up to the seat cut) and take out the entire seat.

For roof designs that call for a tail to support a fascia board, be sure to include more than enough extra wood for the fascia man to work with. Check out what the plans call for on the overhang. If it's a 24-inch overhang, hook your tape in the corner of the birdsmouth and make sure you have at least 30 inches left over. This allows the fascia man to chop off any checks in the ends of the wood.

Using a template.   If you're cutting rafters one at a time, the simplest and quickest way to mark out the cuts for each rafter is by making a template. Use a 2-foot piece of the same material you're cutting the rafters out of. On one end, put your 4 in 12 top cut, and on the other draw the birdsmouth. Look at Figure 12-25.

Take the plumb line on the birdsmouth all the way up to the top edge of the rafter/template, and then square it across the top edge, as in Figure 12-25. This is the reference line you'll match

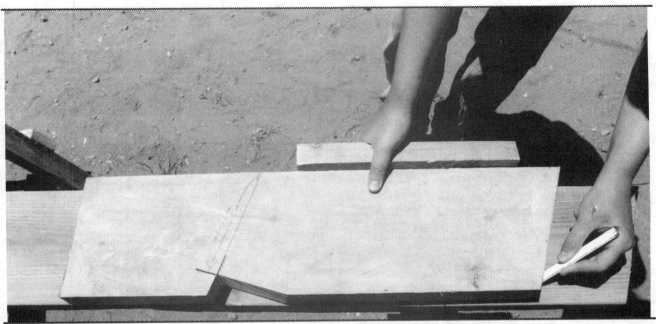

Figure 12-26   When using the jig, start by making your plumb cut line on the top portion of the rafter. If the wood is in good shape with minimal checking along its face, I like to line up the long point of the plumb line with the exact end edge of the wood. Then I can hook my tape on the end of the board to mark the rafter for length. Otherwise I would have to make this plumb cut or set a nail along the top edge at the long point in order to use my tape.

up with the rafter length mark you'll place on the top edge of the rafter. Along the middle section of the top edge of the template, nail on a 2 x 4 block flat and centered on the top edge, as shown. This block will help to keep the top edge of the template in line with the top edge of the rafter you're working with. Just don't cover your reference line with this block!

Now, use the template to mark out the top plumb cut on the rafter material, as shown in Figure 12-26. Cut this mark. Then hook your tape on the long point of this plumb cut and pull out the common rafter length along the top edge of the rafter. (Look back to Figure 12-21.) Make a chicken-foot mark on the top edge at 160⅜ inches. Now place your template on the rafter and line up the reference line on the template with the chicken foot on the rafter (Figure 12-27). Once these two are lined up, you know the birdsmouth is properly located, and you can mark and cut it.

If you have a lot of rafters to cut, this method works really well with a carpenter and a saw working on each end of the rafter. Lay out a few rafters

Figure 12-27   Once you've measured the rafter for length, you can use the jig to mark out the birdsmouth. Line up the reference mark on your jig with the chicken foot you made on the top edge of the rafter that designated its length. Once both lines match up you can mark out the birdsmouth pattern on the jig directly on your rafter and be assured that it's in the right place.

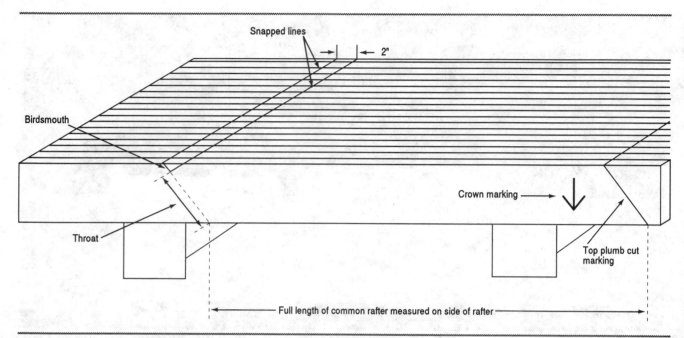

Figure 12-28 To cut rafters in a rack, mark out the two outside edge rafters, then snap lines across the bottom edges from one rafter to the other. When making the cuts, just follow these lines and cut the whole package at once. If the lower throat on the rafter doesn't matter, you can cut the birdsmouth 2 inches wide, measured along the bottom edge of the rafter stock. First make your plumb cut line for the birdsmouth to determine the length of the rafter. Then measure back towards the top of the rafter (along the bottom edge) 2 inches, to the point that the seat squares back towards the birdsmouth plumb line. There are usually support blocks along the sides of the rack, but I left them out of this drawing to keep the cut designations clear. Check out the support blocks that hold up the rafters in Figure 12-17.

on some saw horses. Then have one carpenter mark out the plumb cuts and hold the end of the tape for his buddy who's making the chicken-foot mark and lining up the reference line. Where one person will spend a lot of time carrying his saw back and forth, two people can really fly with this method.

Using a rack. If you have your rafters in a rack, mark out the cuts as I just explained on the two outside boards (Figure 12-28). Then use a chalk line to connect the lines across all the boards in between. Make sure you don't move any boards from end to end before you make your cuts.

Use a circular saw with a swing table set to the proper depth and angle to cut the seat cut. A regular table on a circular saw will make the plumb cut on the birdsmouth. If you're swimming in money (or work) you can buy a circular saw with a dado blade that will cut the whole birdsmouth in one pass. A dado blade is actually a combination of about 13

blades all bolted together. With the proper angle set on the table, it'll make the birdsmouth in one pass. They run around $700.

For the top plumb cut, a chain saw jig will zip through in one pass. If you can't justify the expense for it, the next step down is a suicide blade. With the suicide blade, I like to make one cut all the way down the line with a regular circular saw and blade set at the proper angle. Then I have a bit of a start on the cut and an angle to follow. With a suicide blade, you don't have the luxury of any plumb line to follow when cutting. You just follow the last cut you made. If you make one pass with a regular saw set at full depth and the right angle, you have a line to watch if your angle is starting to change. A suicide blade may tend to move around on you. After a few cuts you aren't exactly sure if you're on the right angle or not. Putting in a preliminary cut in the beginning will help.

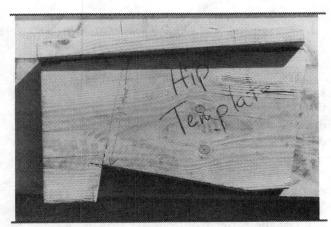

Figure 12-29   Templates really help to speed up the cutting process. The right side of this template has the 4/12 hip cut on it, and the middle has the dropped throat and seat cut worked out on it. With a template, you don't have to work out the angles and seat cuts over and over. Just use the template to trace the parts you need.

■ Cutting the Hips

Hips are usually designed and cut in pairs. If you have a valley of the same span, you could also cut it at the same time. Start by making your top cut. Make certain you use the hip/valley section on your speed square, or 4/17 on your framing square. I recommend making yourself a template (Figure 12-29). Mark out the top, bottom, and length markings for a hip exactly as you did for the common. The only change is that you'll lower the seat a little. More about that later.

When marking out a hip, keep in mind that there are three ways it can be mounted to the ridge. They're illustrated in Figure 12-30. The first has only one cheek cut, the second has two, and the third is resting on the ridge and has only one cheek cut.

The top cut.   To mark out the top cut for the hip design in Figure 12-30a, we'll begin by striking a 4 in 17 mark (Figure 12-13) along its face with the rafter square or the template, making sure the crown is up. Once the plumb cut is on the face, turn the hip on edge. Now mark out the side, or cheek cut. Remember that the cheek for a 4 in 12 hip is 44 degrees. Mark out a 44-degree line on the top edge of the hip, with the long point lining up with the exact location that the plumb cut hits the top edge.

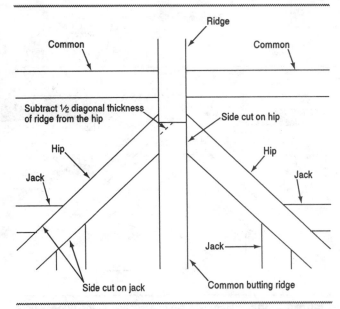

Figure 12-30a   This is a hip intersection at the ridge with a single side or cheek cut. You'll need to subtract half the diagonal thickness of the ridge from its calculated length.

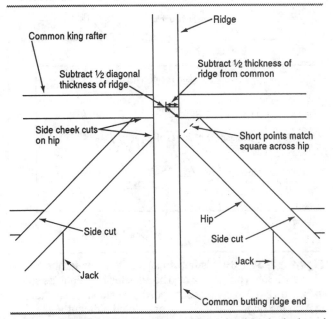

Figure 12-30b   This hip intersection at the ridge is designed with a double side cut. The double cut was needed to make room for the common king rafter. Again, you would subtract half the diagonal thickness of the ridge from the calculated hip length.

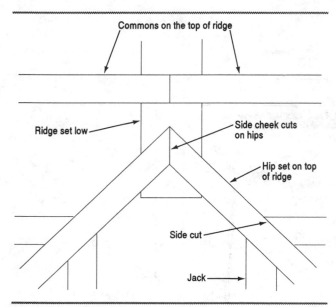

Figure 12-30c Here is a hip intersection at the ridge, where the hips are designed to rest on top of the ridge, with a single side cut on each hip. With this design, you don't have to subtract anything from the calculated hip length to allow for the ridge.

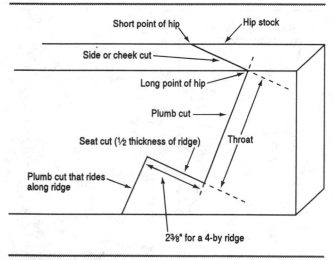

Figure 12-31 Here's how to lay out the cuts for the hips in Figure 12-30b. The throat would be adjusted for height (as shown in Figure 12-32) in comparison with the common rafters that also rest on the ridge. Cut it on the calculated side cut angle. The seat cut will be half the thickness of the ridge material, which, for a 3½-inch-thick ridge, is 2⅜ inches. Lay it out square to the throat. The lower plumb line will be whatever's left over after the seat is laid out. It would also be cut at the calculated side cut angle.

Now you're ready to cut this hip. The hip that opposes this one would be cut the same but with the long point on the opposite face.

For the hip in Figure 12-30b, begin again by striking a 4 in 17 line with the rafter square along the face of the hip. Then turn the hip up on edge and draw a square line across the top edge from the point where the plumb cut intersects with this edge. Turn the hip over one more time and strike another 4 in 17 line on this face with the long point intersecting on the top edge with the square line you just drew. You just transferred the short point over the top edge of the rafter from one face of the hip to the other. These plumb marks represent the short point of the cheek cuts on the hip. Look again at Figure 12-30b. Notice how there are two cheek cuts to allow the hip to fit into the corner made by the ridge and the king rafter. The plumb cuts we just marked out were the two short points on the top edge.

The hips in Figure 12-30c are resting on top of the ridge. When hips are designed this way, you'll notice that the side or cheek cuts are joined together exactly dead center on the ridge. In this case, you don't subtract half the diagonal distance on the ridge from the overall hip length when doing your calculations.

When laying out this top cut, you lay out an upper throat and a seat cut. Look at Figure 12-31. Begin by marking out the plumb cut on the face. Then pull down along this line from the top edge of the hip, the same upper throat as the common rafters for this section. Then establish the length of the seat cut by figuring half the diagonal distance on a 45-degree angle across the top of the ridge. For a 3½-inch-thick ridge, this would be about 2⅜ inches. Draw a 2⅜-inch seat cut and then another plumb cut down from there. See Figure 12-31. The plumb cuts are the long points of 44-degree angled cuts because the top plumb meets its opposing hip, and the bottom plumb will ride at an angle along the face of the ridge.

Dropping the seat. One other adjustment will be needed for this upper throat before you cut. This same adjustment will be done on the bottom cut throat of the hip, so we'll go over that next. Just remember to apply the same adjustment to both the top and bottom throat.

When marking a hip for length, do it along the top edge as you did the commons. The only difference between marking out a common and a hip is in making the throat. This always needs adjusting on a hip, but never on a valley or a common. A valley throat will match the common throats.

When a plane of rafters approach a hip, they should plane in with the middle of the hip. But since they come in contact with the edge of the hip first, and the middle is lower than the edge, the group of jacks will make the roof flare up if you don't make an adjustment here. With a 1½-inch-thick hip, this problem is only slightly noticeable. But with a 3½-inch-thick hip, you'll definitely see the roof flare up at the hips. There are two ways to deal with this. You can take off the edges of the hip by ripping them down at an angle. Or you can shorten the throat (which lowers the whole hip) as shown in Figure 12-32. Lowering the throat (called *dropping the seat*) is more common because it's much easier.

To find the exact distance to lower the hip, first take the thickness of the hip and divide it in half. Let's say you have a 3½-inch-thick hip. Divided in half, that's 1¾ inches. Now make a new birdsmouth plumb line 1¾ inches towards the top cut of the hip, in effect making the hip 1¾ inches shorter. Measure down this new line and mark out your throat height again. Then make your typical seat cut mark.

Now, if you extend the new seat cut back until it intersects with the original wall line, you'll have your dropped seat. But remember this: Don't cut along the second birdsmouth plumb line you drew. This was made only for the purpose of establishing the dropped seat. Cut up the original plumb line and along

the second seat you created, dropping the hip ¼ inch to ¾ inch, depending on the material.

The king rafter. When making the top cut on the hip/valley, some people prefer to leave out the king rafter and make just the 4/12 hip/valley face cut at a straight 44 degrees (Figure 12-30a), instead of adding the second cheek cut for a king rafter (Figure 12-30b). The cheek allows a common against the hip when they both connect at the ridge, and it makes a clean-looking intersection. But you have to add the king rafter if it doesn't fall on layout. I've done it both ways and really don't find any difference.

But there's one situation where there's an advantage to not adding this king rafter and the cheek cut on the hip. When stacking tall steep roofs, extremely long hips (30 feet) may need adjusting to make them plane in with other long hips. A cheek cut locks you in if your common is already in place, and makes the adjustments very difficult.

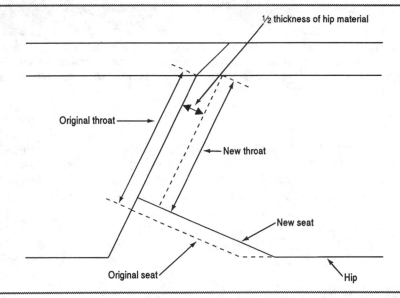

Figure 12-32   To adjust the depth of the throat on a hip, begin by first laying out the birdsmouth so the throat matches the commons. Now find half the thickness of the material being used. If it's a 3½-inch-thick hip, half the thickness is 1¾ inches. Draw a new plumb line, 1¾ inches toward the top of the hip. Now measure down this new plumb line the same length throat that you had before, and make a mark. This is the new seat cut. Mark out this line from the bottom edge of the rafter to the old throat or plumb line. Notice that measuring down the second plumb mark, using the same throat depth, raises the seat cut. When the seat cut rises, the hip lowers. When making the cut, follow the old (or original) plumb line and the new seat cut marking. The second plumb line was only for layout purposes.

Figure 12-33   I like to design my roofs so that the first jack, on all the hips, is always the same length. This makes the jack packages almost all identical, with one more or less depending on the common rafter lengths. This keeps the calculating to a minimum. After you cut your jack packages, keep them in a neat pile or you'll lose a few. These were also stacked on 2 x 4s so the pettibone can get under them and feed them to the framers 20 feet up.

This planing procedure can be hard enough without a common rafter right in your face. With valleys, I almost never cut a double cheek, unless circumstances (such as double rafters exactly set for skylights) make it necessary. And you can't use standard hardware, such as the Simpson MSUL-R series, if you're using a king rafter.

■ Cutting the Jacks

Cutting jacks is a long journey with no shortcuts. One at a time is how it's done. You can save time by working on a wide and tall stack of lumber, marking out six or so at a time. Set up your cut list on a scrap piece of lumber so it's right in front of you. Cut the long ones first so you can use the cut-offs for the short ones.

- Hip jacks get the same seat cut as the commons. Measure them just as you would a common, but to the long or short point of the calculated side cut angle where they meet the hip.

- On opposing jacks, cut their long or short points on opposite sides of the top edge. The top cut would be a 19 degree common 4/12 angle, not a hip/valley angle.

- Give valley jacks a common top plumb cut where they meet the ridge, and a common plumb cut with the calculated side cut angle where they meet the valley.

- Flying hip/valley jacks extend from a hip to a valley that are set close together with no commons between them. Cut parallel common plumb cuts, with parallel calculated side cuts.

Figure 12-34   The package to the left is a complete roof package ready to be taken away by the pettibone. The commons are on the bottom. The hips are banded and placed in the middle and the jacks are banded and placed on the commons. Separate commons for other buildings are racked up to the right. You'll often see packages on tracts banded up. Once they're cut and banded, it's no longer the cutter's responsibility if idiots break the bands and use the wood elsewhere (a common problem on tracts).

On a big job, make sure to keep the packages stacked neat to keep from losing the smaller jacks. On a tract you'll see them banded together so other carpenters don't rip off a few when they need a little backing. Figure 12-33 shows four packages ready to go.

One package takes care of both sides of one hip or valley. By keeping your first jack the same throughout the whole job you'll be saving yourself a world of calculating. All these packages are identical. Figure 12-34 shows the commons, jacks, and ridges all banded together for a particular section stacked and ready for the pettibone to deliver it to some hungry stackers on down the line.

■ Cutting the Ridge Board and Blocks

The only thing missing is your ridge board and common blocks. The ridge board should be long enough to extend a couple of feet past either hip connection. For gable roofs it only needs to be as

long as the building. If your material isn't long enough, splice the ridge by making a 60- to 70-degree cut along the faces. A ridge board is really only a convenient way to attach opposing rafters, so splicing doesn't make it weaker. For lateral shear support, you might strap this joint after it's up. If you're working with a self-supported ridge beam, it's a different story. This is the main support of the roof. *Never* splice it.

The blocks are the last thing you'll cut. The best method is to set someone up on a cross cut saw as shown in Figure 12-35. The blocks will generally be cut from the same material as the rafters, so you can use up all the scrap pieces you have left over from cutting the rafters. If your rafters are set at 16 inches on center, your blocks will be 14½ inches long. If your rafters are set at 24 inches on center, the blocks will be 22½ inches long.

I'm assuming that your rafter material is milled to 1½ inches wide. It's a good idea to check it before you cut 500 blocks. In California they use wood milled in the Northwest, which arrives at the job site

Figure 12-35   Cutting up mountains of common blocks is a long job that's made easier with a cross cut saw. This job had close to 600 14½-inch common blocks.

about 1⁹⁄₁₆ inches wide. In California you'd want to cut your blocks at 14⁷⁄₁₆ inches, or "14 and a half weak" as a framer would call it. By the way, 14½ *weak* is the exact same measurement as 14³⁄₈ *strong*. Some framers never got around to noticing all those little sixteenth lines on their tapes.

The Pay

Roof cutting can pay anywhere from 25 cents to a dollar a square foot of building. If you're piecing out a custom roof, you'll often be bidding on the cutting and stacking together. Some tracts will have roofs that are mostly trusses, but with one section of conventional framing. You may cut the whole tract in a day. This would be the lower end

of the scale. Some framers specialize in cutting and stacking large customs, and inexperienced contractors are only too glad to have their expertise. For a very cut-up roof, they could get top dollar to cut and stack it.

The roof cutter supplies all the necessary tools and calculations to get the job done. The structural roof page of the blueprints will tell you the size and spacing of all hips, valleys, and common rafters throughout.

If you plan on cutting roofs for a living, or if you'll be doing some real complicated irregular roofs, you may need more information than I've provided here. Order a copy of *Roof Framing* from the order form in the back of this book. It's over 400 pages of detailed instruction on figuring cuts for just about every roof you'll encounter in a lifetime.

THIRTEEN

Stacking Conventional Roofs

There's an art to cutting and stacking a conventional roof. If you watch the artist at work, it looks simple. And if you ask the artist, he or she will probably tell you it *is* simple. That is, after the learning is done. The seasoned artist can recognize approaching problems and deal with them before they get out of hand.

One of my philosophies (and it holds doubly true for roofs) is that all work, in the end, is solving a finite set of problems. If you can put aside your fear of the word "problem," you're on your way to being able to foresee the things that are about to go wrong. When you can do that, you can command a high price for your skills.

Here's my point: It takes time and experience to become an artist — and there's no better time than now to begin. Don't be overwhelmed by the thought of cutting or stacking a difficult roof. Just settle in to learn the skills you need.

On a conventional roof, you cut and install each rafter, hip, valley and ridge board on the job site. On a custom job, you might be faced with a very complicated roof design like the one in Figure 13-1. On a tract project, there's usually one person or crew cutting the roof (all rafters, ridges, hips and jack rafters) and a different person or crew *stacking*, or installing, the roof. If it's a small tract, or if times are slow, you'll probably find a single crew cutting and stacking.

Almost all roofs have a ridge board. But notice I said *almost*. Figure 13-2 shows an exposed rafter system that's designed without a ridge. But this is a rarity. Most roof systems aren't exposed, so they're designed with a ridge board to help support the rafter ends and to transfer lateral strength along the top of the roof.

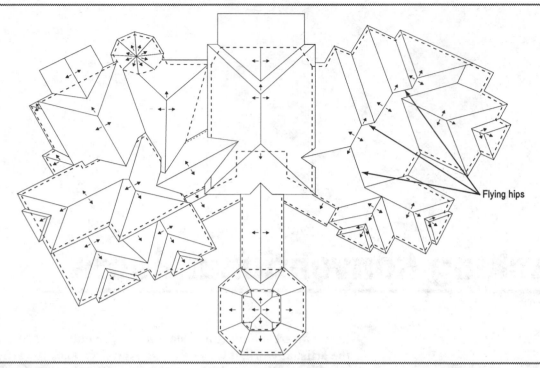

Flying hips

Figure 13-1   This is a view of the roof plan for the roof shown in Figures 11-6 through 11-15. The pitch was 5 in 12. The three flying hips are labeled.

Figure 13-2   These exposed 10 x 10 rafters were designed without a ridge. Three 12-inch long lag bolts installed on the top edge of the rafters hold the plumb cuts securely together. The roof was sheathed with full 2-inch-thick pine so it was plenty strong.

You'll begin by setting the commons and ridges. There are two ways to set a ridge. The first is to calculate the ridge height, then set it at that height. The second is to use four common rafters to hold the ridge where it falls naturally. Setting the ridge with four commons is easiest, when you can do it that way. But sometimes you can't use the rafters to set the ridge. Your only choice then is to calculate the height of the ridge and set it first. We'll start this chapter by calculating the ridge height. It will help you to understand some of the theories behind stacking.

Calculating the Ridge Height

If you have to set the ridge by figuring the height, it's usually because only one side of the ridge has commons, while the other side has something more complicated going on. Look at the ridge

Figure 13-3   What a mess! This is known as a "bastard section." It's stacking with few rules: Make it up as you go, but be sure it all calculates and planes. At this point, all the commons, ridges, and hips have been set. This is a classic example of having to set a ridge by calculation. The ridge to the right has commons only on the right. The calculator gave me the height, and the commons along the only square wall gave me the line.

on the right side of Figure 13-3. You can see that we were able to hang commons on the right side of this ridge, but not the left.

Figure 13-4 shows the same ridge in the middle of the picture. You can see what eventually happened on this roof. We set the ridge at the right height and installed the commons on the right side to make sure it was level and straight. Then we could confidently go on with the more cut-up side.

Most ridges that you set by calculating the height are self-supporting. That is, they're held up with posts in a wall (Figure 13-5) or posts running down to flush beams (Figure 13-6). With either type, there's no need for temporary supports. Just figure the post heights and set the ridges on the braced-off posts. Then you can use the commons to

center the ridge (Figure 13-7). Once the ridge is centered, you can go ahead with the hips, valleys and jack rafters.

Calculating a ridge is a lot like calculating the rafter and hip/valley packages. If you made it through the last chapter, this should be a breeze. Basically, if you know the roof pitch and the width of the building, you can quickly find the height of any ridge in the structure.

Let's use an example. We'll find the ridge height on a building that's 36′6½″ wide, with 97-inch walls and a roof pitch of 5 in 12. To begin with, we need to find the span of the rafters. A rafter spans exactly half of the building width. To get the original width of 36′6½″ into your calculator, convert it into inches (438.5 inches). We'll divide 438.5 in half to find the correct rafter span: 219.25 inches.

Figure 13-4   This is a lot of work just to drain water! The ridge to the right that the stacker is straddling is another calculated ridge. The commons off of it went down to a front wall. The ridge in the middle is the one we talked about in Figure 13-3.

Figure 13-5   This post-supported ridge is held by two posts that run down to the floor. The posts were calculated when the walls were being framed, so they could be built directly into the rake wall.

Figure 13-6   We left this ridge a couple of feet too long on either end in order to accept the hips. At this point they have it set for height and are centering it with a few sets of opposing common rafters. Now is the time to see if the top cuts on the commons work. Always double-check your measurements when calculating ridge heights. The crane has gone home by now, and this glu-lam ridge would be no fun to raise or lower.

Figure 13-7   We set opposing commons on either end of the ridge to make sure it was centered between the walls. Then we rolled the rest of the rafters. These rafters sat on top of the ridge. We gave them a 5½-inch throat on top so we could block with 2 x 6s.

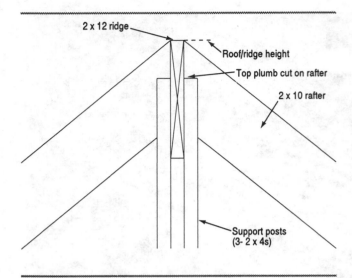

2 x 12 ridge

Roof/ridge height

Top plumb cut on rafter

2 x 10 rafter

Support posts
(3- 2 x 4s)

Figure 13-8 This shows a ridge held up to the top points of two common rafters. In this case the top of the ridge would be the top of the roof. The 2 x 12 ridge is supported by sandwiching three 2 x 4s: one directly under it and one on each side.

Keep in mind that our roof is a 5 in 12. For every 12 inches it moves horizontally, it rises vertically 5 inches. What we need to figure out is how many twelves (or portions thereof) we're moving horizontally from the outside wall to the middle of the room. Once you know that, you need the same number of fives (or portions thereof) going vertically in a 5/12 pitch roof. That gives you the ridge height.

The calculated rafter span is 219.25 inches. How many twelves are in 219.25? Divide 219.25 by 12 and you get 18.27. So there are 18.27 twelves running horizontally. For a 5/12 pitch roof you'll also need 18.27 fives running vertically. Multiply 5 by 18.27 and you get 91.35 inches. That's the distance a 5/12 pitch rafter will rise in a span of 18.27 feet.

■ Add the Throat

But there's one more step. Remember that when the rafter is placed on the wall, it rises vertically above the top plate by the amount of the *throat*, or stand. Try to cut the height of the throat so a standard plumb block will fit in perfectly without

having to rip it down. (More about this later.) You've got to add the height of the throat to the calculated rise to find the height of the ridge above the top plate. With a frieze block this won't matter.

Let's assume the rafters have a 9¼-inch throat, which would allow you to use a 2 x 10 block. You'll add 9¼ inches to 91.35 inches and get 100.6 inches. So the top of the ridge will be 100.6 inches above the top plate of the walls. To find the height of the top of the ridge above the *floor*, just add the wall height:

$$100.6 + 97 = 197.6$$

To review, let's go through the numbers quickly, starting with the width of 438.5 inches:

$$438.5 \div 2 = 219.25$$
$$219.25 \div 12 = 18.27$$
$$18.27 \times 5 = 91.35$$
$$91.35 + 9.25 = 100.6$$
$$100.6 + 97 = 197.6$$

I strongly recommend writing *all* your calculations down on a piece of wood that you plan to keep around, so you have something to refer to when you run into trouble. If the ridge isn't exposed, I like to write my numbers directly on it. Then when it's lifted in place, I have everything I need right there where I need it.

So the ridge is 197.6 inches off the floor. Whether you calculate the ridge height or set it with commons, it will end up the same height. But let me clarify one thing: When I say the ridge is 197.6 inches off the floor, I actually mean that the *top of the roof* will end up 197.6 inches off the floor. If the rafters attach to the ridge as they do in Figure 13-8, your calculation is finished.

■ If Rafters Bear on the Ridge

But what if the rafters bear on the ridge, as they do in Figure 13-9? Obviously, you'll have to set the ridge lower, by the height of the throat on the upper plumb cut on the rafter. So subtract this upper throat to find the height of the ridge. Notice the "top throat measurement" in Figure 13-9.

Most roof cutters match the throat on the top with the throat on the bottom. Remember, the throat on the bottom is the amount of wood left above the birdsmouth, directly in line with the exterior of the building. One good reason to match the throats is that it takes two steps out of your calculations when figuring the ridge height. You don't have to add the lower throat and then subtract the upper throat. Let's do the calculation again with this in mind.

The rafter span was 219.25 divided by 12, or 18.27. Then we multiplied 18.27 by the rise of 5, and got 91.35. Then we added the lower throat (9.25) and got 100.6. If the ridge is set below the rafters, we have to subtract the upper throat. If we cut the upper throat at 9.25, we would end up back at 91.35! Since the two throats cancel each other out, we can eliminate the addition and subtraction.

Setting the Ridge by Height

Now that you know the height of the ridge, you can go ahead and actually set it. If your ridge has supporting posts that are set on the walls, use the height of the ridge above the top plate instead of above the floor. When you're up setting the ridge, you'll usually be walking around on a catwalk or scaffolding that runs down the center of the room. It's easier to measure down to the top plates rather than clear down to the floor.

Begin by finding the center of the span. If you have a 3½-inch-wide ridge, make a mark 1¾ inches off center on both supporting wall top plates. This marks the sides of the supporting posts. Since it's set up 1¾ inches off center, you could also use this line for a temporary brace that would pass along the side of the ridge, then face nail into it.

You'll need a post or temporary support on either end of the ridge. We know the ridge should be 100.6 inches off the top plate. If you cut a support this length, you can nail it flush with the top of the ridge. Then you don't have to mess with the tape when you're up holding the ridge. If the support is resting on a ceiling joist, remember to subtract the thickness of the joist when cutting the support. I

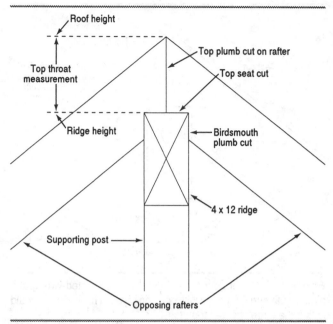

Figure 13-9   Here's an example of a ridge held low with the rafters bearing on top of it. The rafters have a plumb cut as well as a birdsmouth cut. The top throat should match the bottom throat to make calculating easier. In this case, the top of the ridge *is not* the top of the roof.

usually lay a 2 x 4 across the top of the joists and then toenail the support to it. So for 2 x 6 joists, I subtract 7 inches from the support. (That's 5½ inches for the joist and 1½ inches for the 2 x 4.) Then the top of the support ends at 100.6 inches from the bottom of the joists (or top of the wall, since the joists rest on the wall).

Tack up the supports, then lift the ridge and nail it flush with the top of the supports on either end. Have someone steady the ridge while you bring up a common to nail to the wall. Then allow the ridge to float over until its top intersects with the top point of the plumb cut on the rafter. Put up another rafter on the opposite end of the ridge. The ridge should now be in line with the center of the room, and parallel with the wall that the rafters are nailed to. Check that it's level, and make any needed adjustments.

If your plans show posts built into the wall that will support the ridge (as in Figure 13-5), this process is even easier. First locate the center of the

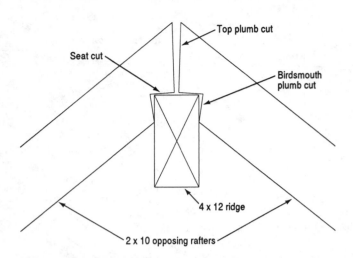

Figure 13-10   If your ridge height was calculated wrong, the birdsmouth and plumb cut will look cockeyed. This is how it would look if the ridge was set too high.

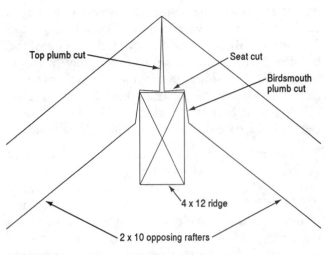

Figure 13-11   Here is how the plumb and birdsmouth would look if you set the ridge too low. It's time to get the calculator out and find out what you did wrong!

room and lay out the size of the posts on the bottom plates. If the posts go from the bottom plate all the way up to the ridge, you'll need to cut out the top plates. Plumb up from your layout marks on the bottom plates and mark out the top plates. A Sawzall works great for making this cut.

We know that the ridge height is 197.6 inches off the floor. Since the post is resting on a bottom plate, subtract 1½ inches from 197.6 to get 196.1 for the post height. Next, because the post supports the ridge, subtract the size of the ridge. If it's a 4 x 12, subtract 11.5 (after you check to make sure it's exactly 11½ inches tall). Let's summarize:

$$197.6 - 1.5 - 11.5 = 184.6$$

If you have any hardware that connects the post to the ridge, make sure you allow for it also. A PC cap takes up about ⅛ inch and a CC takes up about ¼ inch.

If the top and bottom throats on your rafter are the same, you're ready to cut the post. If the top throat is an inch taller, you'll need to subtract an inch from the post height. If it's an inch shorter, you'll need to add an inch to the post.

Once you've measured and cut the post, install it and nail it securely. Then it's just a matter of setting the ridge up on the post. Once it's set, drag up a couple of commons and make sure the ridge is centered and at the right height (Figures 13-6 and 13-7).

■ How to Correct a Mistake in Ridge Height

Look at the top birdsmouth cut where the rafters sit on the ridge. When the rafters are nailed to the top plate of the walls, the plumb cut on the upper birdsmouth of each opposing rafter should be tight against the ridge, if the ridge is centered. If the ridge is set at the proper height, the upper seat cuts will be riding level, and the points of each opposing rafter will meet perfectly, as in Figure 13-9. If the ridge is too high, the seat cuts won't fit right and there will be a gap between the points of the opposing rafters, as in Figure 13-10. If it's too low, the points touch, but the lower part of the plumb cuts will have gaps, as in Figure 13-11. Go through your post and rafter calculations again and find the prob-

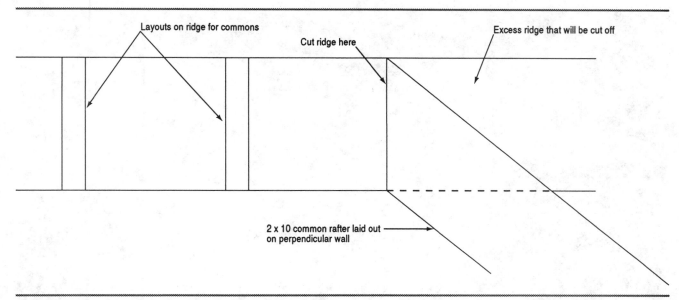

Figure 13-12   To find the point where you should cut the ridge to accept the hips, pull up a common that's laid out directly in line with the end of the ridge. When the top of the common's plumb cut matches the top of the ridge, scribe along the plumb cut onto the ridge. This will be the cut line for the ridge. You can go ahead and nail this common in place.

lem. Recut the commons (if they're wrong) or adjust the post until the commons on both sides of the ridge fit properly.

Setting the Ridge with Commons

To set a ridge using commons to establish its height, you'll need two opposing square walls. First, crown the ridge board. Always put the crown up on ridges. Try to select really straight material for all your ridges!

Before you set the ridge, I recommend that you lay it out as in Figure 13-12. The ceiling joists should be set already so it's only a matter of matching the layout that they've established. Remember that you'll nail the rafters beside the CJs, so keep that in mind when you lay out the position for the first rafter on the ridge. Then, from the first layout, pull 16 inches on center down the length of the ridge. Lay out both sides of the ridge for the opposing rafters.

For most lighter roofs (that is, roofs built with 2-by ridges rather than heavier 4-by material), you'll set the ridge using commons. First establish where the two ends of the ridge will land. For a gable roof, the ridge reaches to the outside ends of the building.

Also check your fascia detail. You might want to leave an extra couple of feet of ridge out past the exterior of the building to support the fascia. If you really want to help out whoever's setting the fascia, detail the end of the ridge to accept the fascia. This detail will change from one job to the next, but it's always easier to cut the ridge end before you set the ridge. Look at the top left corner of Figure 13-13. Notice how the ridge was cut down to a 2 x 4 where it extends out past the exterior wall. This extension of the ridge will help the fascia-setting process tremendously.

■ Splicing Ridge Boards

If you have a long run of rafters, you may need to use more than one length of ridge (Figure 13-13). But it's not enough to just butt two ridges together.

Figure 13-13   If you have a long run of common rafters, you'll probably need to use more than one length of ridge board. It's not enough to just butt the end joints of the ridges together. Instead, cut a 24-inch-long angle on each ridge to help transfer the load and stress through each ridge board. Also notice how the left ridge was detailed on its left end to accept the fascia overhang. We ripped it down to a 2 x 4, as was detailed on the plans. The rafter tails were also cut to accept the soffit detail.

Professional roof cutters always make at least a 24-inch angle cut on both ridge ends to help transfer the load. Make a 24-inch mark, as shown in Figure 13-14, and then cut away.

■ Setting the Ridge on a Hip Roof

On a hip roof, the ridge stops in from each end by a distance of half the span. When I'm cutting a hip roof, I like to leave the ridge a couple of feet long on either end, and cut it to length right before I bring the hips up to it. Then there's no risk of setting a ridge that's too short for the hips to reach.

Setting the ridge with four commons can be done with two people, but having three is ideal. Start out by leaning up four commons on the out-side of the building — two opposing rafters on each side of the ridge. Have two people stand on the CJs or scaffolding at either end of the ridge to hold the top ends of two opposing rafters as they're passed up to them. Then you can nail the bottom of each rafter to the top plate of the wall, toenailing down into the wall and face nailing into the ceiling joist. If you have hips to go up, nail these first two rafters a few layouts away from the hip connection to the ridge. And make sure you get the rafter nailed well, or it'll blow out when you're setting the ridge.

Now you have two sets of opposing rafters nailed at their bottoms and steadied by two people, as in Figure 13-15. The two helpers are keeping the tops together so the rafters hold each other up.

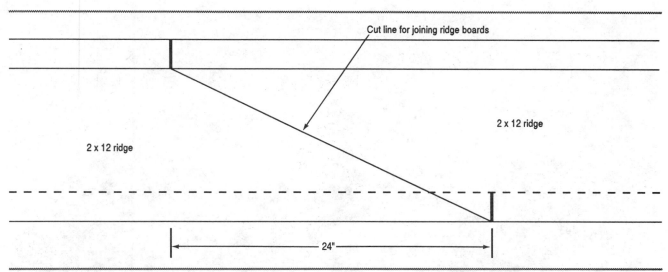

Figure 13-14 When joining two ridges, make the cut at least 24 inches long. Always put the crowns up when figuring this cut.

Figure 13-15 When setting a ridge with commons, it's best to have three people. These two carpenters are each steadying the top plumb cuts of two opposing rafters. A third carpenter has passed up the commons and is down on the top plates nailing the commons securely to the exterior walls.

Figure 13-16   Now they're sliding one side of the ridge between the commons and steadying it.

Now slide the ridge up under the top cuts of the rafters until it's flush with the tops of all the rafters, as shown in Figures 13-16 and 13-17. If you nailed the bottoms well, the rafters should pretty much support the ridge at this point. Nail the tops of the rafters flush with the top of the ridge on their proper layout.

Finally, check the ridge with a level. When a ridge is set with four commons, it's only as level as the walls that the commons are riding on. And the walls are only as level as the slab. If it isn't level, you may have to beat one end down a little. If that doesn't work, you may have to force the other side up with a 2 x 4 under the ridge and on top of a joist.

You don't want to raise or lower the ridge so much that the top plumb cuts have large gaps. Then again, you don't want to spend all day recutting each rafter. You'll need to find a happy medium between going up on one end and down on the other. Whatever you decide, do it now! Don't wait until you have all the commons up before you check the ridge for level. It won't budge at this point. When it's level, put up a temporary sway brace or the whole thing might fall flat. I like to go through the whole roof and get all my ridges set before I fill in the hips and valleys.

■ Installing Hip Rafters

Once the ridge is braced up, you can install the hips. To find the exact point on the ridge that the hips will intersect, pull up the common rafter that's

Figure 13-17   Then they slide the other side of the ridge into place. Once the top of the ridge matches the top of the plumb cuts on the commons, nail the commons on their layouts.

laid out exactly in the middle of the wall that's perpendicular to the ridge. Since the ridge is in the center of the room, this layout will line up perfectly with the end of the ridge. Put the bottom of the common on this layout, and run the top along the excess ridge until the top of the common matches the top of the ridge. Look again at Figure 13-12. Then scribe the end of the common along the ridge to mark the cut-off point. The point where the top of the common intersects with the ridge is the exact place where the middle of the hips will meet the ridge.

Pull the hips up and nail the bottoms flush with the outside corners of the building. With both hips nailed at the bottom, check to see how they meet the ridge. They should match so that their long points meet the ridge equally, as shown in Figure 13-18. If they don't meet equally (as in Figure 13-19) you'll need to persuade one hip with a toe-nail or two so that it lines up with its partner.

Some roof cutters cut their hips with *double cheek* cuts on top. See Figure 13-20. This makes a nice clean connection at the ridge. With a set of hips cut this way, mark the end cut on the ridge with a common but don't nail up the common. It would just be in the way at this point. The end cut on the ridge is the point where the middle intersection of the cheek cuts on the hips will land.

If the hips rest on top of the ridge, they'll have a top seat cut and a 45-degree angled top plumb cut and a 45-degree angled birdsmouth plumb cut, as shown in Figure 13-21. Nail the bottom wall seat

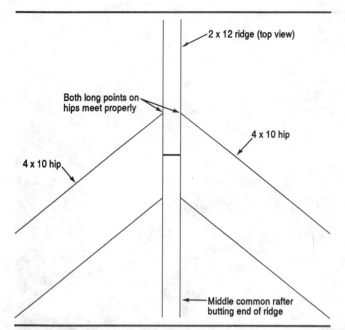

Figure 13-18 When two hips approach a ridge, their long points should match across the top edge of the ridge. The center of the hips should line up with the point where the common meets the ridge.

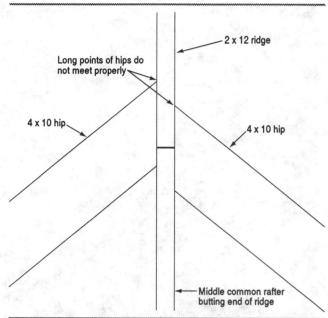

Figure 13-19 If the long points of the hip don't line up across the ridge, you'll have to shift them until they do. Otherwise nothing is going to plane properly.

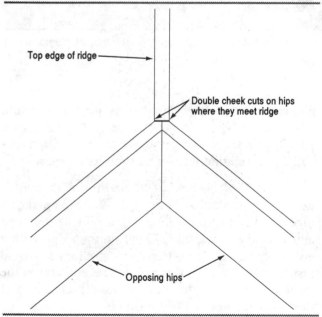

Figure 13-20 Hips cut with double cheek cuts will butt into the ridge. It's a nice clean connection. Find the end cut on the ridge with a common, like we did in Figure 13-12, but don't nail up the common. Just use it to find the point where the cheek cuts on the hips intersect with the ridge.

cuts in place and then set the tops on the ridge. The framers are doing this in Figure 13-22. The 45-degree plumb cuts of each hip should meet cleanly. If they don't, you have one of three problems: the building is out of square, the ridge is at the wrong height, or the hips are cut wrong. Check everything until you have a solution.

■ Installing Flying Hips

Sometimes you'll find a hip detailed on the plans that has a top ridge connection but doesn't go all the way down to the wall for its lower connection. Instead, its bottom is supported by another ridge that's set lower. This is called a *flying hip*, because it seems to float from one ridge to another. Look at the roof plan shown back in Figure 13-1. There are three flying hips shown on the right side of the plan.

If there's another full length hip that leaves from the same top point and goes down to a wall, install it first. Then you'll have a great line to sight

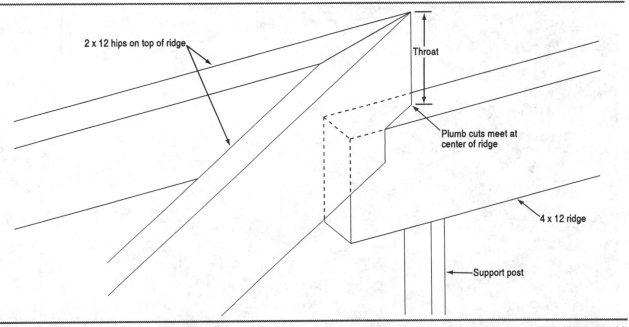

Figure 13-21   If the common rafters are designed to sit on top of the ridge, the hips will also have to sit on top. They'll be cut with a 45-degree angled top plumb cut, a seat cut, and a 45-degree angled birdsmouth plumb cut. The top plumb cut will not match the plumb cut on the common. It will be adjusted so the common will plane with the center of the hip, and not the edge.

Figure 13-22   After this hip was set in place, they stepped back and sighted it with the common rafters behind it. The rafters and the hip should all be in one smooth line. After both hips were set, they cut off the excess ridge beyond the post.

Figure 13-23   In this section, the ridge has been set, along with two hips in the background and a valley in the foreground. The commons on the left are being installed now. The framer got the ridge height from two commons near the hips, and then leveled it out. The rest of the commons on the left put the ridge in line. Notice the hips and stacks of commons and jacks piled on the ceiling joists waiting to go up.

the flyer with. Next you'll have to install the lower ridge that supports the bottom of the flyer. See Figure 13-23. The flyer will also plane with the lower commons on the right side of this ridge.

Calculating the length for a flyer can be a bit tricky. Cutters often guess at the length, then cut it long so the stackers have room to adjust it when they're hanging it. The stackers temporarily hang it and then step back and sight it with the rest of the roof. Then if it doesn't work they can cut it down until it sights nicely.

If you have the lower ridge up and the full-length hip installed, you can use strings to find the length of the flyer. You'll need two people. Look at Figure 13-24. Start your string at the far end of the lower ridge. Set a nail directly in its center, and then pull the string off in the direction of the full-length hip that's up (you can see its top edge in the lower left corner of Figure 13-24). Have one person stand at the other end of the ridge (the end that will receive the lower connection of the flyer). This person will watch for the string to land directly in the center of this end of the ridge. The second person is holding the end of the string over the top of the full-length hip.

Have the string nailed to the far end of the lower ridge, running parallel over its top, and held at its other end on top of the full-length hip. The first thing you'll notice is that as you move the string off center of the ridge and up the hip, it rises above the

Figure 13-24 A flying hip takes off from where the valley left off in the last photo, and connects with a regular hip (in foreground) on a ridge set earlier. A flying hip is a hip that goes from one ridge to another, with no wall connection. Notice how the hip extends the plane of the commons that attach to the lower ridge. Now you can set the jacks, being careful not to bend the hip. A small pop-out has also been stacked to the right. It will eventually California fill onto the commons.

end of the lower ridge that your partner has his eye on. If you move it down the hip, it will just skim the top of the ridge and plane in with the tops of the commons shown on the right side of Figure 13-24. (This assumes the roof is cut properly and the slab is decent.)

You might have a quarter-inch gap, but that's fine for a span this large. The trick is to get the string lying directly down the center of the ridge and then mark this point on the far hip. This will be the length of the flyer.

Once you have a mark on the full-length hip, measure down from the long point on the hip to the long point of the mark you made with the string. Now cut another hip this length for the flying hip. Of course, you'll have to cut the top and bottom 45-degree angles opposite the full-length hip because this flyer will oppose the hip you measured from.

Strings can be wonderful tools when stacking a roof. If you aren't sure if something is working, just drag a string off to an accompanying plane somewhere close by and see how it rides. Soon you'll develop a good eye for spotting problems. But there's always a place where you'll need to pull a string.

Checking and Filling In the Skeleton

Nail the bottom seat cuts of the valleys on their designated layouts on the walls. While a hip will always be attached to the outside corner junction of two walls, a valley will be attached to an inside corner junction. Look at the valley in Figure 13-23. After the bottom is secure, simply swing the valley

Figure 13-25   The hip this carpenter is writing on is the flyer from the last photo. It connected on top like any other hip. You can see that once the skeleton is up, you can sight how all the hips, commons, and valleys plane with one another.

into the ridge and nail the top. Wherever it meets the ridge is where it belongs. Then check that it sights with the commons and hips (Figure 13-25).

Once you've set all your ridges, hips and valleys, it's important to check everything, either with your eye (Figure 13-26) or using strings. Now's the time to adjust anything that doesn't plane. If the walls weren't built square, some funny things are going to happen to the roof. Sometimes it takes a few tricks to make it all plane. Never be afraid to adjust a hip or valley for your eye even if it defies your calculator.

When you sight a roof's skeleton, you're making sure that the angle of every common matches the angle of any nearby hip or valley. Look closely at Figure 13-25. If you stand back a little and match the top edge of a common with a hip or valley in the distance, you'll see if the top edges

of each are matching. You can't be too close to the common that you're sighting over or you won't be able to accurately sight it along its whole length. This is very important. You need to compare its top as well as its bottom with other roof members in the distance.

■ Correcting the Mistakes

Sometimes, when you set two hips to the ridge, their points don't match across the ridge. After you persuade one to get it to match with the other, step back and sight it with a common. Maybe the persuading caused the top edge to rise so that it doesn't sight or plane with the commons. Do both the hips match in length? Were they cut right? Usually it's a problem with the slab being out of level. Some-

Figure 13-26   It's a good idea to sight even a run as simple as this, to make sure it all planes.

times you'll need to pull the nails on the wall connection of the hip and let it slide out past the wall a little. Try this and then sight it with a common. Is it better?

Remember, if it's really out of line now, it will be visible when all the jacks are on and the roof is sheeted. Right now you're only looking at two lines. Later you'll see the whole plane of the roof sweep up or down. I know this from experience — I've messed up my share of roofs!

If the ridges are all set level and the plumb cuts on the commons fit correctly along the ridge, leave them alone. Make any adjustments to the hips and valleys. Double-check your calculations for the length before you adjust them. You can't really change any top cuts on a hip or valley except by making them shorter.

You'll have two choices for the lower wall cuts. You can make the plumb cut that rides against the wall longer, so that the hip or valley will rise up on the ridge when you pull it closer to the ridge. Or you can push the hip or valley out past the wall, leaving a gap between the original plumb cut and the wall (if it will be buried). Both of these will cause the hip or valley to change its plane and affect how it sights with the rest of the skeleton.

Don't be afraid to pull the nails on something you've already hung up, even though you're sure it calculates right. I've hung a skeleton in two days, and then spent a week pulling nails, burning up the keys on my calculator, and adjusting to the concrete man's mess until it all sighted straight. That's when you get really good at working a calculator!

Figure 13-27   When setting commons, always work from one side of the ridge to the other. Set a few rafters, then jump to the other side and set the opposing rafters. Otherwise you could get a ridge that looks like this. Too many commons were set on the right side, which caused the ridge to bend. Notice how the tops of the rafters on the left were beaten silly to try and fix this dilemma. Sometimes you'll need to lift the ridge a little to straighten it. The weight of all the rafters has probably caused the ridge to drop a little.

■ Filling In the Rest of the Rafters

Once you're satisfied with the skeleton, you can fill in the rest of the roof. When you're filling in the commons, be sure to get two toenails into the top plate and three face nails into the ceiling joist if possible. Keep an eye on the ridge. Sometimes it tends to bow one way or another from the pressure

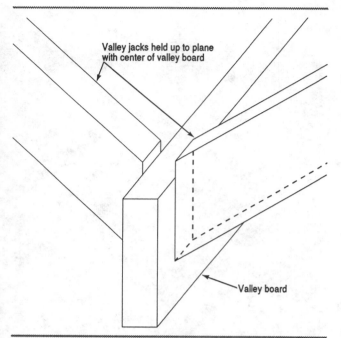

Figure 13-28   When attaching the valley jacks to the valley, be sure to hold them up a bit so they plane into the center of the valley. If you have trouble sighting this, use your speed square along the top edge of the jack and find where it planes into the center of the valley.

of the rafters, like it did in Figure 13-27. Don't fill in one whole side of a ridge and then jump over and fill in all the opposing rafters. Your ridge will almost certainly be bowed out of line. It's much better to put up a few on one side and then fill in the opposing rafters on the other side.

Be extra careful when setting the jack rafters. Bent hips and valleys are very visible. Nail a few on one side of the hip and then move to the other side, progressively working up the hip from side to side. If it's a very long hip you may want to put in some temporary braces to hold it straight.

Jack rafters are cut on a progression that makes them land right on layout along the hip or valley, as long as the wall layout is 16 or 24 inches on center. If your CJs are on layout, the top cut on the jacks should fall nicely against the hip or valley. If they're not, you're going to lose that 16- or 24-inch on center layout.

Figure 13-29   This is a simple California fill. The glu-lam ridge was one continuous piece. Notice the valley was a 2 x 12 laid flat on the sheathing. The 2 x 12 was wide enough to support the complementary cut of the jacks that rested on it.

If you run into this problem, either the building isn't square or the joist layout is off. There isn't much you can do about an out-of-square building except recut your jacks so they work. If the joists aren't on layout, it's probably easier to pull their nails and fix them instead of cutting the jacks.

When you hang jacks on a valley, you'll need to hold them up a little so that their top edges plane in with the center of the valley. See Figure 13-28. This will allow the bottom edge of the plywood to plane in with the center of the valley.

■ California Fills

California fills on conventional roofs can be extensive (Figure 13-29) or just small pop-outs (Figure 13-30). Either way, they're framed just as I described in Chapter 11. Begin by sheathing the roof surface that the fill rests on. Then use strings to find the valley lines.

On conventional roofs you'll usually use the same size rafters as the commons. In Figure 13-29 the California fill jacks are 2 x 10s, which match the commons on that roof. A complementary cut on a 2 x 10 is going to be pretty long, so I suggest you use a 2 x 12 for the valley board that's nailed to the sheathing, or double up a few 2 x 6s.

When you raise the ridge for the common section, it's a good idea to leave it long enough to catch the California fill section. If you look closely at Figure 13-29, you'll notice that the glu-lam ridge was left long. It was also cut with the complementary angle on its end before it was raised. This cut would have been a real bear to do while it was up.

Figure 13-30   A small pop-out such as this can really take some time. By putting a 1-foot cantilever in the room, the architect is adding a little character to an otherwise flat wall and roof line.

On some houses the architects like to break up long wall lines with small bay windows or pop-outs in the room that cantilever out past the exterior wall. These need small roofs that are most easily treated as California fills. Check out Figure 13-30. By measuring the width of the walls, you can figure the length of the hips that rest on the sheathing. Notice the small section of ridge that levels off the ends of the top hip connection. The length of hip that rests on the sheathing had to be ripped down so it planed with the end that sat on the corners of the walls. All the jacks also had to be ripped down to the appropriate height. A little fill like this can take as much time to construct as the larger fill shown in Figure 13-29.

■ Radius Roofs

Another specialty item you might run across is a radius or round roof. These tend to look a lot more complicated than they are. To begin with, you'll have a radius wall. In the roof in Figure 13-31, there was a radius section that formed off of two straight walls. The straight section was framed first, and the ridge left long at the end, just like we do for a hip attachment. The commons end exactly where the straight section of wall ends and the radius begins.

When laying out the rafters on the radius top plate, I like to begin my layout with the center rafter that's directly in line with the ridge. Then I pull a 16-inch on center layout both ways off of it. This center layout will have a common raised on it. Pull up a common and rest the top plumb cut beside the ridge and scribe the plumb cut just like we did in Figure 13-12. Make sure that the top of the plumb cut matches the top of the ridge. After you mark the ridge, cut it and then nail up this common.

After installing the first common, put up another common at every third layout mark. Look closely at Figure 13-31. The top cut gets a cheek cut

Figure 13-31   A round roof is actually very easy to stack. The two common rafters that meet at the end of the ridge are located where the radius wall dies into the two straight walls on either end. Another common should fit directly in line with the ridge, down to the midway point around the radius top plate. Once these three commons are established, you can just fill in the rest. To support the whole thing, there's a post set from the end of the ridge down to a flush-set beam.

at an angle that depends on how it meets the group at the ridge. You'll need to adjust each top cut accordingly.

Now the problem is how to fit all the top cuts cleanly into a quickly-shrinking space at the ridge. The best solution is to install a cross block between the standing commons. Notice the blocks about a third of the way down from the top cuts of the commons. This block is strong enough to support the weight and give you plenty of room to install the remaining rafters. The bottom cut remains the same, but you'll need to take a measurement to establish the new lengths.

■ Installing Blocking

Once the rafters are all rolled, they need to be blocked. Check whether you'll need a frieze block (Figure 13-32) or a plumb block (Figure 13-33).

Sometimes the height of the ceiling joist is more than the throat cut on the rafter. If so, a small corner of the CJ will be standing above the rafter along the exterior wall line. For instance, if you had a 5½-inch throat cut on the rafter and a 7½-inch ceiling joist nailed along its side, you'd have 2 inches of ceiling joist running above the plane of the roof. You'd want to trim this 2 inches off before you did any blocking. If you do it first, you can do it with a circular saw. If you wait until later, you'll have to use a reciprocating saw, which will take ten times longer. Cut the CJs along the slope of the roof as pictured in Figure 13-34.

When you nail the blocks, face nail through the rafter into the block with three 16d nails. Don't worry about nailing the back side yet. Just go down the line and face nail one side of each block until you have all the blocks in one section up. Then turn around and back nail each block in the row. This is quicker and safer because you aren't spinning

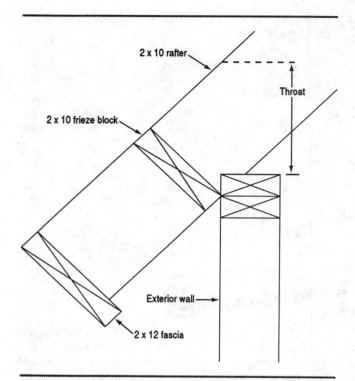

Figure 13-32   A square-set block is commonly called a *frieze* block. The back of the block is set against the outside of the exterior wall.

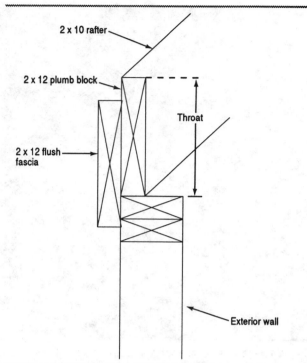

Figure 13-33   A plumb-set block is set in line with the outside edge of the exterior wall. Cut the throat on the rafter to the height of standard material. If you don't, you'll have to rip down all your material before you can cut your blocks — a formidable task.

around to nail each block. When you back nail, keep your nails on the back of the previously-nailed block so the nails aren't exposed and visible from the ground. If they're on the exposed side of the previous block, they'll rust and bleed in the first rainstorm.

If your rafters rest on the ridge, you'll also have to block between each rafter. The block should be full height so it reaches from the ridge it rests on to the top of the rafter. That provides for edge nailing for the plywood. If the top throat on the rafter wasn't cut to accept a common-size block (5½ for a 2 x 6 or 7½ for a 2 x 8, for example) you'll have to rip down material to the appropriate size.

Also check the plans to find out what kind of posts are required under the ridges, hips and valleys to keep them from sagging. Notice all the supporting posts in Figure 13-35. Make sure you send any

supporting posts down to bearing walls when possible. There will often be flush beams installed for the sole purpose of supporting roof members. Use 4 x 4s or three 2 x 4s. If you have a 2-by ridge, valley or hip to support, use one 2 x 4 directly under the member and one on each side of it that run up onto the member and face nail into it. You can see this sandwich effect in Figure 13-8.

The Pay

Stacking conventional roofs will pay from 20 cents to a dollar or more a foot. This can range from $400 for one tract house to $10,000 to cut and stack a large, cut-up design like the one in Figure 13-1 at the beginning of the chapter.

Figure 13-34   If your joists are taller than the throat on the rafter, you'll need to cut off the amount that protrudes above the rafter.

Figure 13-35   When the skeleton is up and you've sighted it for straightness, you can put in the supporting posts. Some posts will go down to flush-set beams and others will go down to top plates or ceiling joists. The plans will give you locations and detail any metal straps you'll need.

Fascia Board, False Tails, and Stuccoed Eaves

After stacking the rafters (but before the roof is sheathed), you need to apply a fascia to give the roof an attractive finish and help define the building's style. A typical ranch-style house usually has a rough-sawn fascia board on the rafter tails and along the gable ends. A Spanish-style building might have corbeled false tails, while a Spanish Mediterranean building could have false tails with some sort of stucco cover. The more distinguished French Villa look often incorporates crown molding with other fine wood to create a layered look.

By far the most popular is the standard resawn fascia board. It's easy to apply, economical, and blends well with any style or pitch of roof. Figure 14-1 shows a typical roof with the fascia hung and ready for sheathing.

Fascia Board

Most fascia is a rough-sawn board that's one size wider than the rafter tails. Depending on where you're doing your building, most fascia is spruce, fir, or pine. Spruce is very popular because it's lighter and easier to handle than fir or pine. The mills take a normal piece of S4S (smooth four sides) wood, and run it through the saw again using a blade that creates a rough finish. Because they only resaw a certain amount of wood at a time, it's important to use wood from the same batch throughout the whole house. Discrepancies from one cutting day to the next can cause you a world of headaches.

Figure 14-1   This house has typical resawn fascia boards set square to the rafter tails. Notice the outlookers set 32 inches on center along the gables. Also notice the 2 x 4 block set in line with the frieze blocks and spanning from the gable truss to the barge rafter. This house is now ready for the sheathing crew.

The mills might have the saw set up to cut a little deeper into the face of the wood than it did yesterday. When you try to match a 45-degree face cut on two boards with a quarter-inch size discrepancy, you'll have a gap either on the front side or the back. And remember, this is finish work that won't get buried by drywall or stucco. You just can't leave behind quarter-inch gaps when you walk off the job — not if you expect to keep your reputation intact.

If the board is only resawn on one side and one edge, keep the rough face out and the rough edge down. It's more economical for a mill to only resaw the edges that are visible, instead of all four edges.

You also want to try to keep the crown up. If the board was carelessly resawn, the rough edge will be on the crown. Then you'll have to decide

between the crown up or the rough edge down. If the board is terribly crowned and you're working on the back corner of a building where it won't be too noticeable, you could probably get away with putting a smooth side down. But always clear a decision like this with your foreman.

Fascia board that's nailed to the end of the rafter tails and along the gable ends should have an equal overhang all the way around. If the rafter tails extend 24 inches past the exterior wall line, the gable ends usually also have a 24-inch overhang. Some jobs will vary and have a smaller overhang along the gable ends. Sometimes you'll cut off the rafter tails and nail the fascia flush against the building. The style of the house (and the taste of the architect) will determine how the fascia goes. Since the rafter tails are usually left long, you'll have to cut them back to create the right size overhang.

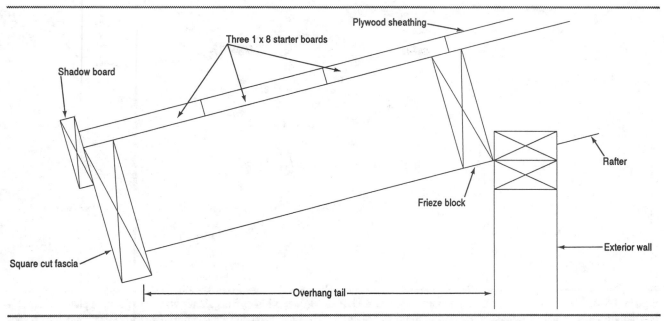

Figure 14-2   This is a typical square cut fascia setup. The three pieces of starter board extend from the end of the fascia to the middle of the frieze block. The shadow board in this example was set 1 inch above the roof sheathing to support the front edge of the first piece of concrete roof tile. Each row of roof tile laps on the previous row laid below it. If we don't raise this shadow board an inch, the first row of tile will dip forward and look strange compared to all the other rows. With asphalt shingle roofing, you'd mount the shadow board flush with the top of the fascia.

On the gable ends, you'll set *outlookers*, which are 2 x 4s set flat that run out from the second rafter or truss, over the top of the gable rafter or truss, and out to the fascia. Check out the gable ends in Figure 14-1 and you'll see the flat 2 x 4s. They support the middle of the fascia at the gable ends.

■ Planning the Fascia

To begin, start by spreading out all the wood. Determine what length of board each run will need and lean them against the building accordingly. Long runs might require a few boards that are spliced together with 45-degree angles. Always try to use boards that are a little longer than the run. Then you have some leeway to cut off any split or checked ends.

The fascia on the gable ends are typically supported with 2 x 4 outlookers. You can splice if you have to, but it's much better to use a full-length board because you'll have only 1½ inches to tighten

up any splices. There's a special splice that you can use, as I'll explain later. But for now, as you're spreading the material, keep in mind that you want to try to use a full-length board on the gable ends.

Finding the overhang. Next you have to find the exact overhang of the fascia. The plans will call out the approximate overhang. Typical plans call for resawn 1 x 8 T&G to sheath the overhang from the fascia to the middle of the frieze blocks. See Figure 14-2. This is commonly known as *starter board*. Starter board gives the underside of the overhang a nice finished look.

The starter board ends exactly in the middle of the frieze block, and the plywood starts where the starter board ends. This gives both materials a border to edge nail into. You want the overhang from the outside edge of the fascia to the middle of the frieze block to be exact, so the framer putting up the starter board can lay down a whole number of boards without having to rip any down. Using the approximate overhang from the plans, you can cal-

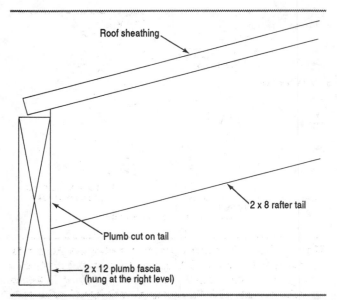

Figure 14-3   A plumb cut fascia hangs parallel to the walls. This figure shows a plumb cut fascia applied at the right height. The front edge of the fascia should be in line with the plane of the top of the rafters. This allows the plywood sheathing to lie flat.

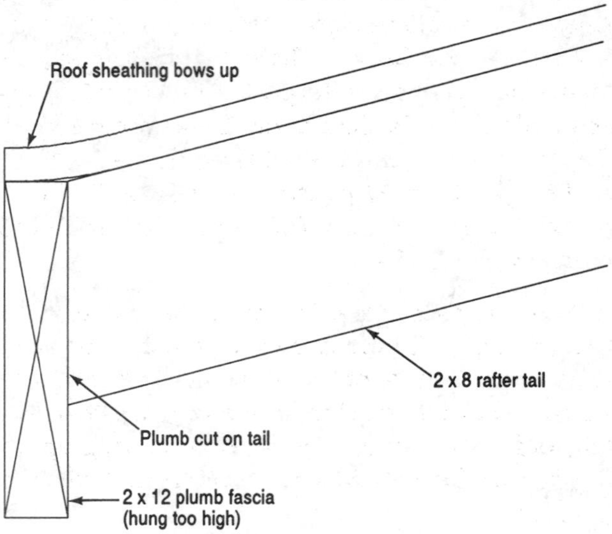

Figure 14-4   This shows how *not* to hang a plumb cut fascia. By lining up the top edge of the back of the fascia with the top edge of the plumb cut on the rafter tail, you'll cause the plywood to bow up at the end.

culate how many whole pieces of 1 x 8 you need to span it. That gives you the exact overhang, which is the point where you'll cut the rafter tails.

The house in most of the photos in this chapter scaled a 24-inch overhang on the plans. We joined three pieces of starter board together. Their total width was 20½ inches, which was close enough for the builder. So the overhang along the eaves was a *three board overhang*. We subtracted 1½ inches for the fascia and cut off the tails at 19 inches from the center of the frieze block, or 18¼ inches from the face of the frieze block. (We'll use that 18¼-inch measurement in the examples that follow.)

Square or plumb cut rafter tails. You'll either cut the rafter tails square or plumb at this point, depending on how the plans show the fascia hanging. A *square cut* is just a 90-degree cut on the rafter tail. Figure 14-2 shows a square cut fascia. That will leave the fascia board square with the rafter, but at an angle to the side of the building. If the plans show the fascia mounted plumb with the building wall, you'll mark a *plumb cut* that matches the slope of

the roof. See Figure 14-3. For a 4 in 12 pitch roof, you'll cut a 4/12 (18.5 degree) angle on the rafter tail to make the fascia hang plumb.

When you hang a fascia on a square cut tail, you'll want to hold it flush with the top edge of the rafter tail. The roof sheathing or starter board will then be applied to the outside edge of the fascia and edge nailed into the fascia.

A plumb cut tail is different. If you attached the back edge of the fascia even with the top of the plumb cut, the front edge of the fascia would be above the plane of the top edge of the rafters. When the sheeting crew applied the plywood or starter board, it would hit the front edge of the fascia and cause a severe bow. See Figure 14-4.

This problem is relieved by holding the fascia down a little bit until the front edge of the fascia planes with the top of the rafters, as shown in Figure 14-3. Most framers find this distance by holding the edge of their speed square along the top of the rafter tail, and then slowly sliding the fascia up until the outside face edge hits the underside of the square. Then they nail it up.

Barge rafters. On the gable ends, the fascia is known as the *barge rafter*. It's supported by flat 2 x 4 outlookers that are "let-in" (or cut flush) 1½ inches from the top of the gable truss (or rafter) and end nailed into the first truss. Leave them long so they cantilever out, then cut them the same length as the rafter tails. Place these outlookers 32 inches on center. You can see the outlookers and barge rafters in Figure 14-1.

Like any job, organization is the real key. The trick is to get all your material (fascia, outlookers, nails, a few lengths of 2 x 4, and your circular saw) within reach so you don't have to come down off the roof once you get up there. Traveling up and down a ladder will waste your time and wear out your legs.

Once you've spread all the fascia board, cut the outlookers to approximate lengths (always going a little long to be safe). From the ground take an approximate measurement from the second rafter or truss in to a point about 30 inches out past the exterior wall. Then cut enough 2 x 4s to lay out outlookers 32 inches on center up the gable truss. Tack a nail into one edge and hang them on the bottom chord of the truss they'll be nailed to. Then you can reach them when you need them. If you can get all the material within reach, you'll save a lot of time. Once you've spread all your fascia and outlookers, pick a run where you'll start the installation.

■ Installing the Fascia

Pull up the first fascia board and lay it face down on top of the tails, about 3 inches above the point where you'll cut them. Set two nails into the top edge of the rafter tails at either end of the fascia so it won't slide off. Now you have a walkway. Mark out the two end rafter tails of your first run for the cut length (ours was 18¼ inches) from the face of the frieze block, and snap a line along the tails. At the chalk line, mark each tail for either a plumb cut or a square cut.

Marking and cutting the rafter tails. If you're marking out plumb cuts, use a Squangle or make a template so that you don't have to find the angle on

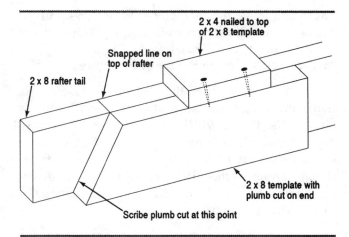

Figure 14-5   To make easy work of marking out plumb cuts, try making this template. Cut the appropriate plumb angle on a scrap piece of rafter material, then nail a short piece of 2 x 4 down flat to its top edge. The 2 x 4 will keep the top edges of the template and the rafter tail in line. Once you have the top edge of the template's plumb cut in line with the snapped line on the rafter tail, you can scribe a line along the plumb cut and onto the rafter tail.

your speed square for each tail. A Squangle is a square that you can set at any angle with a small wing nut. Once you have the angle set, it's just a matter of laying the tool on the top edge of the rafter tail and lining up the blade with the snapped line. It'll hold the same angle as long as you don't loosen the wing nut.

If you don't own a Squangle you can make a simple template out of a piece of material that's the same width as the rafter boards. See Figure 14-5. Use about a 12-inch piece of this material. Mark out your plumb cut on the board and then cut it. Now tack on a small piece of 2 x 4 centered along the top edge. This 2 x 4 will keep the top of the template in line with the top of the rafter tail. Line up the top of the plumb cut on the template with the snapped line on the rafter tail and then mark the tail. Once you have a line of rafter tails marked out, go ahead and cut them.

Cutting the fascia. With the tails cut, you can cut the fascia itself. When cutting miters on fascia, always cut on the back side of the board. A circular saw will pull up and shatter the wood on the face and leave the other side clean. By cutting on the back, you leave the face clean. When you pull up the fascia and use it as a walkway, lay it face down so it's ready to cut. That way you're also walking on the back so you're not marking up the face.

You want to miter the end of the fascia that hangs out to meet the barge at 45 degrees. The back side of the fascia will be the short point for this miter. Once it's cut, make a mark on the back side of the fascia (at 18¼ inches in our example) from the *short* point of the miter. Set the board back down on the tails so that the outside of the gable truss is in line with this mark.

If you're hanging the fascia alone, you'll have to hold and nail it somewhere in the middle at first, so you'll need a reference mark to go by in the middle. If your fascia is laid down with the correct overhang (the 18¼ lines up with the exterior wall line), you can mark a reference line. Go to the middle of the fascia board and make a line along the side of a rafter tail where you'll be balancing and nailing at first. When you start to nail the board, this line is your guide so you know the fascia will leave the correct overhang on the end for the barge.

Of course, a piecework framer will always figure out a way to do a job alone if possible. Why not do it alone and have your buddy taking care of something else? But if you do have two people, one on either end of the fascia, all you'll need is the end 18¼-inch mark. One person can watch this line and call out when it lines up with the outside of the gable end.

With the overhang end mitered and the fascia set along the tails in its exact location, you can locate the miter cut on the other end of the first fascia board. First, decide where you'll join this board to the next, in the center of one of the rafter tails on down the line. If you tried to join the fascias in a bay between rafters, you'd have no tail to nail into and snug down the miters. The tail ends hide most of the back side of the miter while they give you a lot of wood to nail into.

Locate the tail that's nearest the end of your fascia stock as it lies exactly in place, after you've cut off any splits in the end of the lumber. You

always want to cut the first board you put up with the short point of the miter on the outside finish face, and the long point on the back. This allows you to lap the next fascia mitered end *over* the first. Otherwise you'll have to slide the second miter *under* the first, which is difficult to make work cleanly. Once you've marked it, carefully make a slow, clean cut at 45 degrees. Remember, this is finish work. Make every cut a beauty!

Hanging the fascia at gable ends. When hanging fascia, you'll always want to use galvanized 16d nails. Begin by starting a nail at the reference line in the middle of the fascia board. Then slowly slide the board over the edge of the tails, balancing it in the middle with one hand. When the reference line is where it belongs, nail the fascia into the rafter tail with the other hand. Set the nail about three-quarters of the way in. Next, go down and make sure the 18¼-inch mark lines up with the outside of the gable truss. If it does, nail it also.

Finally, go check the other end. Did the miter land as you planned it, over the rafter tail for a good connection with the next piece? If everything worked out, finish nailing off the fascia. If you have 2 x 4 rafter tails, get two 16d galvanized into each tail. If you have 2 x 6 or larger, try to get three nails through the face of the fascia and into each rafter tail.

With the first piece up, move down the line to the piece that joins with it. Always work in one direction. Pull up the next piece and lay it down as a walkway. Put a 45-degree miter on the end that will join with the board you just nailed. On this piece, the short point of the miter will be on the back side and the long point will be on the face side.

Now you need to find out what happens on the other end of this second piece. Does it join another gable at 45 degrees? If so, you can take a measurement to establish its length. Since you want to cut on the back side of the second piece of fascia, tack a nail into the long point of the fascia that's already hung (which should be its back side as well), and measure to the outside edge of the gable truss. Since you measured from the long point of the installed fascia, this measurement will be the short point of the second fascia.

For the other end, remember to add 18¼ inches (for the gable overhang) to the short point of the miter on this end. This measurement is called "short to short," since we're cutting from one short point to another.

After you cut the second piece to length, lay it exactly in place and make a middle reference line for hanging purposes. Slide the board over the edge, tack the middle nail, and then check the ends for accuracy. Make sure the mitered lap is tight before you sink any nails.

Hanging the second piece from the middle and getting the joining miter cuts to match can sometimes be a real struggle. If you aren't exact with your reference marks (within ⅛ inch), you'll have a gap in your miter joint. If you have trouble here, try this trick:

1. Once your fascia is cut and exactly in place, go to the end opposite the middle miter joint and sink a nail in the top edge of the fascia exactly in line with the middle of the gable truss or rafter. Sink the nail about half way in and then bend it over at a 90-degree angle. This nail will serve as a "hang nail."

2. Go to the middle miter joint and set a few nails into the face of the board. The hang nail is easily disturbed, so you want to set the nails now and create as little movement as possible later.

3. Once you have some nails ready, slide the fascia over the edge so it's in place. The hang nail will hold one end of the fascia while you support the other end and make your way down to the middle miter. You can move the fascia lengthwise a half inch or so to line up the miter, but not much more than that. The hang nail holding the gable end is precarious, so be careful.

4. Line up the miters and then nail them together. You may want to leave the long point of the miter lapping a little long over the short point of the underside board. When you drive the nails in, the miter tends to move diagonally. If you anticipate this movement by leaving the miter a little long, it will end up perfect after you've nailed it. *Don't miss when swinging your hammer or the shock will send the fascia down.* Make sure there's no one working below you if you try this trick.

5. When you have the mitered end secure, go down to the gable end and put a tack in there. Before you nail it all off, kneel down and see if the top edges of the two fascias are in line with one another. You might need to raise or lower this gable end a fraction above or below the tops of the rafter tails in order to keep everything looking straight.

When you're satisfied, nail it up. Always be careful when nailing fascia. Anytime you miss, that big scar in the wood will be there forever. If you miss with a waffle-headed hammer, it's going to look like an elephant's been walking on your wood. Do clean work and keep your good reputation.

Hanging the fascia at valley corners. If the second piece of fascia ends on an inside valley corner, you can run it a foot or so past this corner, then cut it off later when the fascia it meets is in place. You'll get a more precise inside joint if you have the other board up when you make the cut.

On truss roofs, the valley corners are rarely framed up by the stacking crew. This is because most valleys on truss roofs are formed by California fills rather than actual valley boards. The piece of valley overhang that ends up exposed and in line with the roof's valley is applied by the fascia crew, because the fascia ends up supporting one end of the valley.

Look closely at Figure 14-6. After the fascia was installed, the valley was attached to it and run up to the corner of the supporting wall. Then small jacks and the missing frieze blocks went in.

Figure 14-6   Here's one way to fill in a valley for a truss roof. Since the actual valley is supported by the fascia, it must be done by the fascia person, after the fascia is hung.

For conventionally-cut roof valleys, you'll need to make the double miter plumb cut on the valley rafter. Look at the valley rafter where it attaches to the fascia in Figure 14-6. You'll notice the plumb cut was made with two 45-degree miters that meet in the middle of the valley. A conventional roof will usually have a much bigger valley board than this small 2 x 4 that was installed for the truss valley.

Since the valley board is already installed on a conventional roof, you find the location of this cut when you snap out each overhang on top of the rafter tails. The lines intersect on top and in the center of the valley. This intersection establishes the top of the plumb cuts.

When making the cut for two fascia boards that meet on an inside corner such as this, don't try to miter the corner. It comes out cleaner if you just butt the joints.

■ Installing Barge Rafters and Outlookers

After the horizontal fascia boards are up, you can hang the gable end barge rafters. To begin, you need to set the outlookers in place. Go to the top of the gable end and pull a layout 32 inches on center down the truss. Cut a notch 3½ inches wide and 1½ inches deep on each layout. Drop the outlookers into these slots and end nail them into the first truss in. Sight the gable truss to make sure it's straight and then nail the outlookers into it, making any adjustments necessary to make the gable straight. These flat 2 x 4 outlookers will support the barge rafter along its middle.

The bottom end of the barge is supported by the lower fascia that we let hang out 18¼ inches. The top is supported by the excess ridge. On convention-

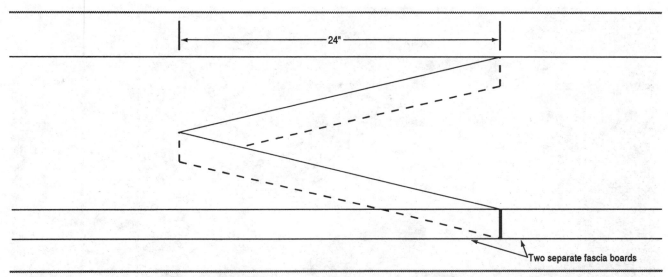

Figure 14-7   If you don't have material long enough for your barge rafter to stretch from the ridge down to the lower connection, use a connection like this to join two pieces together. A regular 45-degree miter is impossible to tighten up. This type of joint is very strong and will withstand the weather over the years. After it's hung and painted, it's nearly invisible.

ally-cut roofs, the stackers will almost always leave the ridge long to provide upper support for the barges. You'll need to cut this back to the proper size for your overhang.

If the ridge wasn't left long to support the fascia, nail on a ridge block to support the two plumb cuts of the barge rafters. This ridge block should be one size smaller than the fascia. If you have 2 x 8 fascia, use a 2 x 6 block. Cut it 18¼ inches long and toenail it to the ridge. This block will accept nails from the two top cuts of the barges. On truss roofs you'll never have an extended ridge at the gable end, so you'll always need to add this block.

Hopefully you'll have fascia material long enough to reach from the ridge down to the lower fascia connection. If not, you'll have to splice two boards. You can't do this with miters because you don't have anything to attach your miters to. To splice the two boards, make a joint like the one pictured in Figure 14-7. You'll cut a 24-inch long "V" into the board on the left and a 24-inch long point that matches the "V" in the other board. When joining the two together, force the point into the "V," so that it's tight. You might want to put a little wood glue on before you join them.

After they're together, drive in four nails, two on the top and two on the bottom. Always manufacture this connection on the ground, and then lean the board up to mount it on the gable end. It's strong enough that you don't have to worry about handling it. After this joint is painted, it becomes nearly invisible.

Lean up your barge rafter material along the lower run. Then take these steps:

- Snap a line on the outlookers from the ridge block to the short point of the fascia that cantilevers out 18¼ inches from the gable end. Cut off all the outlookers at this line except the top one. You'll need it for support.

- Lay the barge on the outlookers and make the 45-degree miter cut that will meet the lower fascia. Turn the fascia on edge (the way it will hang) and move it slowly to the outer edge of the outlookers.

- Working from the bottom, pull the board over all the outlookers except the last one, which remains uncut (Figure 14-8). You're holding the bottom half of the barge, and the last uncut outlooker is supporting the upper half.

Figure 14-8 When hanging the barges, start by nailing the top of the lower miter cut together. Notice how the top outlooker was left uncut in order to support the top of the fascia.

Balancing it on this outlooker, and working on the lower connection, make sure the long points of the miters meet closely at the top edge. The bottom edge will gap open because the fascia is way too high on top while it's resting on the last outlooker. Don't sink the nail, just set it in far enough to hold it.

- Take the saw up to the ridge, and, holding the fascia, cut off the last outlooker along the snapped line (Figure 14-9). Slide the fascia down into place and tack it into the last outlooker, letting the excess fascia run wild at the ridge.

- Once the top is secure, go back down to the lower miter cut and check that it's a good fit. Use 16d nails to tighten it up. Then work your way up the barge, sending two nails into each outlooker.

- Make a plumb top cut in line with the middle of the top block and then nail the plumb cut into the ridge block.

- Hang the opposite barge at the lower miter, then scribe its ridge along the barge that's up, so the cuts are identical. Once you get good at hanging barges you'll be able to make the top plumb cuts for both boards by eye at the same time, letting the blade mark the first board for you as you cut the second one (Figure 14-10).

- Once the barge is up you'll need to install a 2 x 4 block in line with the frieze blocks, from the gable truss out to the barge. This block will support the starter board when it changes direction from along the rafter tails to up the gable end. The block will also hide the joint that occurs when the starter board changes

Figure 14-9   Hold the fascia with one hand while you cut the last outlooker on the line snapped earlier.

Figure 14-10   If you have a real good eye, sight the top cut. Otherwise, scribe the back side and watch the blade as you slowly make your cut.

Figure 14-11   The first step to hanging fascia is to cut the tails along an evenly-snapped line. The hip must be cut with two 45-degree angles. Because a circular saw's table only adjusts for angle cuts in one direction, one of these cuts must be made from the bottom up, as shown. Notice how the board to be hung is laid on top of the rafters and used as a walkway.

direction. Cut the block at 18¼ inches and keep it in line with the run of frieze or plumb blocks. Nail it to the gable and the inside face of the fascia.

■ Installing Fascia on Hip Sections

To hang fascia on hip sections, you can let the two side pieces run long and snap a line from one fascia to the other, along the tops of the rafter tails. Then make all your cuts and hang the piece between them. Or you can work your way around the hip as shown in Figures 14-11 through 14-17. This keeps you from having to drag the saw back and forth.

To begin, snap all the rafter tails on the hip section for overhang length at once. The hips will have two miter cuts that meet at a point in the center of the hip. By snapping all the tails at once, you'll mark out both corners. Cut the tails for length and then cut the miters on the hip. Because the table on

a circular saw only adjusts to a 45-degree angle in one direction, you'll need to make one of the miter cuts on the hip from underneath. If you're really good with a saw and comfortable with heights, try cutting from underneath, as shown in Figure 14-11.

Once the tails are cut, you can make the cut on the back end of the fascia. In our example, the fascia died into the roof. This calls for the complementary angle cut. We had to hang the fascia with this cut a few inches above the roof sheathing so the roofers could slide their waterproofing paper under it. The material was a little short so we made a miter cut here (Figure 14-12) and then put a small piece in later. You'll cut the outside miter after the board is hung, so run out a little extra at this point. Make a reference mark, swing the board over the edge, and nail it in the middle (Figure 14-13).

Cut the outside corner hip next. This ends up being a plumb cut, even though the rafters are cut square. The bottom edge of the face is the short point and the top edge is the long point. Find the

Figure 14-12   Cut a 45-degree angle where the board will be lapped before you hang the fascia.

Figure 14-13   Holding the board in the middle, lower it over the edge and nail it to the rafter tail that's next to your reference line.

Figure 14-14   Cut the corner of the fascia so the short point ends up right in line with the long point on the hip. Make sure the hip is strong before you go monkeying out on it!

middle of the hip and transfer this mark on a 45-degree angle to the face edge of the fascia. Set your saw at 45 degrees and make the cut (holding on tight with your knees)! Check that you followed the line and then clean up the cut if you need to (Figure 14-14).

For the front piece, use a hang nail to hold the board temporarily and scribe the back side of the cut along the piece that's already hung (Figure 14-15). You'll be marking the long point of the first piece of fascia along the back of the second piece. But the back of the second piece wants to be a short point! What you're doing is matching angles so the outside corner is perfect. After you scribe the board, pull it up and transfer the marks to the front of the board so you're cutting the long point you scribed. That's what's happening in Figure 14-16.

When the end is cut, use a hang nail to help put it up. In Figure 14-17, the hang nail is on the hip on the right. Then you can work on the other hip,

Figure 14-15   A hang nail was set into the top edge of the right side of the fascia and bent over. This helps support the board while the carpenter makes a scribe mark from the board on the left to the board on the right.

Figure 14-16   Because the long point was scribed on the back side, it needs to be transferred around to the front side, and then cut.

Figure 14-17   Using the hang nail on the right side again, the corner miters are nailed together, and then the rest of the tails are nailed off.

Figure 14-18   Hold false tails down 1 inch below the top of the rafters so the 1½-inch starter board planes in with the ½-inch roof sheathing. False tails are usually put alongside every other rafter.

making sure that the miters go together nicely. Notice that the fascia runs long on the right hip. The framer will move over to this corner and fix it up.

The technique I described here is for hanging board alone. If you're new to fascia, it might be best to get a partner and help each other set it. Then you don't need the reference lines. If you work in pairs, have one person spreading material as the other starts in one corner hanging fascia. Once the material is spread, the first person will start hanging outlookers. When he's done, he'll go to the opposite corner of the building and start hanging fascia.

There are tricks for every level of framing that let you work alone. But unless you're pretty skilled, it can be a frustrating experience. It takes some practice to master these tricks and use them to make your job easier — and more profitable.

■ Hanging Shadow Board

Sometimes you'll have to hang *shadow board*, a smaller board over the fascia that's held flush to or above the top of the fascia. It may hold up the first row of roof tile, or just add a shadow line along the fascia. There's a shadow board in Figure 14-2.

Shadow board is usually 1 x 4 material. Most piece workers will spread the 1 x 4 at the same time they're spreading the fascia so they have it right there when they need it. You don't want to drag your saw all the way around the building twice.

You'll usually miter all butt joints on the shadow board. For square-set fascia, set the top edge of the 1 x 4 even with the top of the fascia. For

Figure 14-19   Make sure the false tails are laid out symmetrically. The middle bays can vary up to 3 or 4 inches if necessary, but the bays on either side of a hip or valley should be almost exact.

plumb-set fascia, you'll need to hold the top edge down so that it lines up with the bottom of the sheathing. Otherwise you'll bow the sheathing.

False Tails

False rafter tails are used to give the impression of size. But to keep the effect from being too heavy, they're usually placed every other bay. They're cut from 4-by or 6-by material to exaggerate the overhang. The ends of the tails are usually dressed up by corbeling or rounding them. You can do this on the job site or order them already cut from the lumberyard.

Since false tails are usually installed without a birdsmouth cut, it's important to adjust the throat of the rafters *before* they're cut to allow the different roof sheathings to plane.

Because the tails are placed every other bay, the span is too large for a 1 x 8 starter board to provide sturdy support for the roofing. You'll need to use 2 x 6 or 2 x 8 resawn starter board. And remember to install the tails 1 inch lower than the top of the rafters (Figure 14-18). Then, when the 2 x 6 starter board is mounted on the tails, it will plane in with ½-inch plywood sheathing on top of the rafters. This is very important if the roofing material will be light enough that it won't hide any change in elevation in the roof sheathing.

The most important step in installing false tails is the layout. A shoddy, poorly-planned layout is visible for miles. The tails near the hips and valleys have to be synchronized and spaced exactly. Notice how the tails on both sides of the hip tail in Figure 14-19 are laid out identically. That will only happen if you plan the false tail positions while you're laying out the ceiling joists and rafters. In other words, false tail layout starts during the ceiling joist layout.

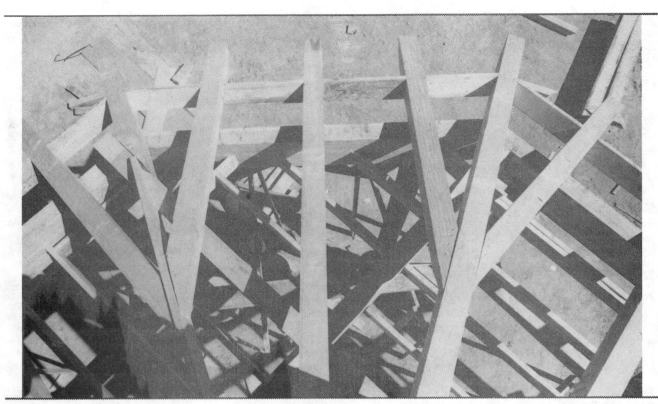

Figure 14-20   An awful lot of work goes into hanging false tails, as you can see by this corner alone. Not only are they heavy, they also require a beam saw to cut them. The frieze blocks were held out here because the whole structure received thickened arches around its perimeter.

Start in the center of the long stretches of rafters, and lay out a bay exactly in the center. Then lay out bays of equal size at either end, near the hips. Keep everything symmetrical by matching the layout on both sides of hips and valley.

Hip and valley sections will give you the most trouble. Notice the two different hips in Figure 14-20. The one on the right worked nicely in comparison to the one on the left. In a small section like this we decided to just use the false tail as the hip. Notice how the hip on the left ran into trouble. The beam that it's attached to was a supporting beam for the roof and couldn't be moved. The only choice was to frame around it. By using straightedges and strings we got it to plane and fall in line.

Keep in mind that the false tails need to be directly in line — none of them hanging out farther than the rest. Use a string to make sure. Mount the two end tails, and run a string from one to the other. Now, as you mount the tails in between, make sure they just touch, but don't push into, the string. You also need to make sure that one tail isn't cocked up in relation to its neighbors. When you mount your string, get it on the upper and outermost point of the tail. Then when you bring each tail to the line, you're checking for both length and height.

Stuccoed Eaves

The simplest overhang to apply is the stuccoed eave. This can vary from a single band to an elaborate layering of bands and a curvature of the rafter tail to give the eave a slight roundness. The most popular stuccoed eaves involve the layering of successively smaller bands (Figure 14-21). It's quick and easy to apply because you don't need to make any mitered cuts. You can even use leftover scrap wood as long as it's fairly straight.

Builders will often try to imitate a crown molding for their eaves. They'll cut ¾-inch plywood into any number of designs to achieve a slight curve in the eave. After the plywood template is attached to each rafter tail, the stucco crew staples the stucco wire and paper to the plywood and forms the mud to the design. This can be a very labor-intensive process, when you start cutting up several hundred templates.

There's an easier way to get this rounded look. Put a slight curve on the lower part of the rafter tail and then mount some bands plumb on the top part of the rafter tail and on the wall just below the rafter tail (Figure 14-22). Then you can achieve the curve on the tail with two slight cuts, rather than an actual curved cut. Make the curve cuts and the plumb cut

Figure 14-21 This is a typical banded eave. Notice that there's no need for mitered corners, or clean ridge cuts. Just nail it down good, keep the reveals constant, and don't look back!

Figure 14-22 This stuccoed eave has a slight curve in the center of it. Instead of using plywood templates to achieve the curve, try putting a couple of slight cuts on the rafter tails below the plumb cut. This will give the stucco person enough backing to create the desired curve between the upper and lower bands.

Figure 14-23   This is how the detail in Figure 14-22 looks when it's finished. The parapet wall received a similar detail.

on the ground, when you're cutting the rafters. After the roof is stacked, run a 2 x 6 and 2 x 4 band on the rafter tail as well as along the wall. The stucco crew can staple the paper and wire along the bands and accentuate the curve when they mud. You can see the final product in Figure 14-23. Notice the parapet wall got the same treatment. In my experience, this is a lot easier than installing plywood templates.

The Pay

On a piecework basis, fascia board, false tails, or stuccoed eaves are applied by a different crew from the one that stacked the roof. The piece price pays per foot of length of overhang applied. False tails are usually set by an hourly crew if they're framing the whole building.

Fascia will pay per foot of length for the entire amount of fascia installed, with the price depending on its size, the size of the rafter tail ends (a long overhang with 2 x 4 tails is somewhat scary to stand out on, but extra pay will usually soothe my nerves) and whether or not there's a shadow board. I've made $1.00 per linear foot of board for a 2 x 8 board with a 1 x 4 shadow board. I've also made $2.00 a foot for 2 x 6 without shadow board because the job was so cut up. It was all short pieces with a lot of cuts. It didn't have any long runs at all, where you can earn some quick money and make up for the cut-up areas.

Roof Sheathing

S heathing the roof is one of the most basic steps in framing a house. With only the bare minimum of tools and a day of training, a person can apply sheathing on nearly any roof. Once you have the starter board or the first row of sheets down, it's relatively safe and easy to walk around. That makes sheathing a piece of cake compared to the constant danger that stackers face while they're walking the top plate!

In sheathing, like all framing jobs, there are tricks you can use to save time. One of the biggest timesavers in sheathing is getting the material loaded for easy access. You can build a rack against the building, you can load the sheathing on the rafters, or you can pass it up by hand, one piece at a time. By far the best way is to have it loaded on racks with a crane or pettibone. If you get a job piecing out the roof sheathing, the material should be loaded on the roof, or on racks that you can grab from easily when you're up on the roof.

The majority of roofs are sheathed with ½-inch plywood. If there's an exposed overhang on the roof, sheath the rafter tails from the fascia to the frieze block with something that looks nicer than plywood. The usual choice is 1 x 8 tongue-and-groove pine *starter board*. If the fascia crew did a good job, the overhang was calculated so you can fit an exact number of boards between the outside edge of the fascia and the center of the frieze block (Figure 15-1). This saves on labor, as well as material, because you don't have to rip and fill any areas. Usually it's a three- or four-board overhang with the last board against the building ending exactly in the middle of the frieze block. See Figure 15-2. If it's running up a gable end, it would end in the middle of the gable truss (or last rafter). This allows edge support and nailing for the plywood.

Figure 15-1 This shows a typical truss eave with a three-board overhang. At the bottom of the gable, the lower boards run through. There's a block that runs in line with the frieze blocks and spans from the gable truss to the fascia. This is the point where the starter board changes direction. The block allows nailing for both edges, while hiding the joint. Notice how the material is run long on the bottom row. Always lay as much material as you can before cutting.

Starter Board

On piecework jobs, you'll spread your own starter board. Start by getting all your material leaning up against the fascia around the perimeter of the building. Once you get up on the roof, you won't want to keep coming down for more material.

Starter board may be smooth or rough sawn. The builder will let you know which side he prefers down. The "down" side would be the exposed side that's viewed from underneath. If your material is surfaced smooth on both sides, put the best side down. Sometimes you'll have one side that's surfaced smooth and also has an angle cut along its joint edge. When you join the edges of each piece with the next board with the angles facing each other, you'll end up with an inverted "V" groove rather than a straight, butted joint. See Figure 15-2. This is common on starter board that's smooth surfaced on both sides.

If you're using 1 x 8 starter board, use 8d nails to tack it into place. If you're using 2 x 8, use 16d nails. The 2 x 8 boards are a little tougher to join together, so you may have to use a few toenails to persuade the joints together.

When applying the starter board, start by covering the rafter tails along the bottom runs. Lay down the outside piece and tack it with an 8d nail so one edge splits a rafter tail and any excess runs wild over the fascia (Figure 15-1). Keep the bottom edge of this first piece flush with the outside edge of the fascia. Slide two more into the grooves and nail them up when they're tight. A good fascia crew will have added a 2 x 4 block on edge, running in line with the frieze blocks. It's nailed from the outside gable to the gable fascia (or barge rafter as it's usually called). If not, you'll need to put one in.

This block hides the joint where the starter boards change direction from the horizontal lower runs to run up the gable on top of the 2 x 4 outlookers. The lower run of starter board will split this block down the middle, leaving the upper half of the block for the gable pieces to nail onto. It's good practice to lay down as much material as you can and then make all the cuts at once.

If you're sheathing over false tails, you'll need to find out where the contractor wants the starter board to begin. Some want it out to the edge of the tail. Others may want it back a little so that a few inches of the tail are exposed beyond the starter board, as shown in Figure 15-3.

Here's something you should be aware of: If you use a 1-by starter board, an 8d nail driven straight through the starter board may poke through the bottom of the outlookers. Avoid this by angling

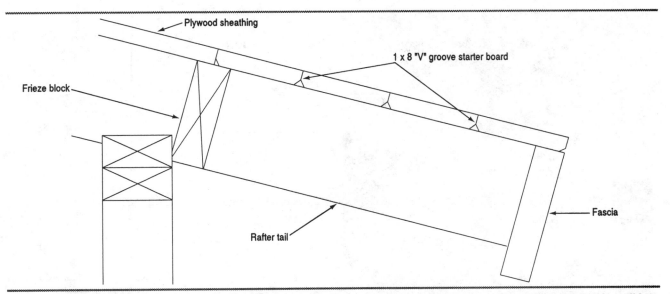

Figure 15-2   Starter board is laid from the front edge of the fascia to the middle of the frieze block. The plywood then takes off from there. Starter board that's smooth on both faces will usually have a "V" groove milled into it. This groove goes down (as shown) so you can see it from underneath.

Figure 15-3   Place a whole number of 2 x 6 starter boards on false tails. Some contractors like running the starter board right out to the front edge of the tails.

Figure 15-4 The starter board and plywood will follow all the changes from flat runs of commons, over hips and into valleys. A job with a lot of hips and valleys will take a lot longer than a straight gable roof.

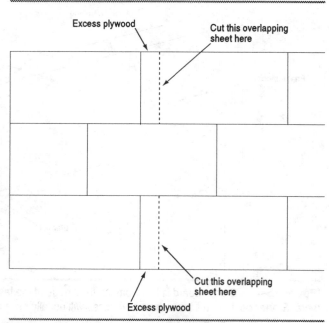

Figure 15-5 When laying down sheets, always stagger the vertical joints from one horizontal row to the next. If you have an unexpected change in rafter layout, try lapping the sheet you're laying down back over the sheet you just laid. This allows you to nail the front edge of the plywood on a rafter so you can go on with sheathing and not be slowed down by having to make a cut. Snap a line on the section that's lapped back over the first sheet at the point where the edge of the first sheet (the sheet underneath) ends. Later, when you're making your cuts, you can simply cut along this line and the cut edge will fall right in place.

your nails on the gable ends so the nail tips are still in the wood. Exposed tips can rust and bleed, causing big stains in the wood after it's painted. Make sure you mark out all the tails and outlookers for the nailer. If there are any vent frieze blocks, you might want to mark them also so the nailer won't shine any nails down through the vent holes. (The expressions "shine any nails" or "shiners" refer to exposed nails that miss the support wood they were intended for.) Vent blocks with shiners are almost impossible to deal with. Always mark out the vent blocks to avoid this problem.

Inspectors always look for a disproportional number of shiners. They expect to see a few, but you'll get called for row upon row. Shiners are acceptable in a buried attic space, but along exposed overhangs you'll have to pull them out.

When you lay down the starter board, you'll want to keep it following the rafter tails perpendicularly throughout the whole building. Follow each change in direction no matter how small. See Figure 15-4.

Sheathing the Roof

Once the starter board is down, you're ready to lay the plywood. Start by snapping a line parallel to the fascia, 48 inches from the center of the frieze block (the end of the starter board) up the rafter. This will ensure that the first row of plywood is straight. Lay down the first row, going by the

Figure 15-6 Starting at the bottom, lay out the plywood sheets. Snap a line 48 inches up the rafters from the starter board. Notice how the plywood laps over the edges of the starter board. Just lay down your sheets and ignore the overlap. Then, when all the sheets are down, do your cutting. Always mark out the rafters as they pass under the sheets. This task is indispensable come nailing time.

snapped line and *not* the starter board. Make sure each edge of the plywood ends in the middle of a rafter. Sometimes you'll need to push or pull the rafters a little to get the plywood bearing on them correctly.

If the plywood won't reach the next rafter because of a change in layout, lap it back over the last piece until it does end in the middle of a rafter. See Figure 15-5. Then snap a line where the plywood is lapping over the last sheet. But don't make this cut until you have all the sheets down. Whenever you make a cut on the roof, you create sawdust. Walking on a roof that has a fine layer of sawdust on it is a little like walking on ball bearings. It's not too bad on a 4 in 12, but it gets a little unnerving on a 6 in

12. Anything steeper than that becomes downright suicidal! An organized sheather lays all the sheets he can before he even takes his saw out of the truck.

When starting a new row, start with a half sheet so your vertical joints will jog (Figure 15-5). You don't want any vertical joints to match up from one horizontal row to the next. When a sheet is down, get a nail in each corner and then proceed. When you have a whole row down, mark out the rafters where they pass under the sheet, as shown in Figure 15-6. Always mark out the rafters or you'll have a heck of a time when it comes time to nail off the roof.

Follow the hips and valleys, keeping the grain of the plywood perpendicular to the rafters. See Figure 15-4. Lay all the whole sheets that you can

Figure 15-7   This exposed beam ceiling received 2 x 8 T&G sheathing on its lid. Notice how the ends are left long and then cut off all at once. This ended up being a great surface to work off of when we stacked the roof.

and then use your cutoffs to fill in the hip and valley triangles. This will help save on wood. Jobs with a lot of hips and valleys are going to take much longer than jobs with straight gable runs.

The sheathers who know how to make money may work ten hours a day on a big roof, just laying down sheets. The next day they pull out their saws and cut for two hours, using their cutoffs for spots they've planned in advance. If you have a big cutoff at the end of a line and you know it's at least big enough to start your next row, you can leave the space open, knowing that the cutoff will fit in when you get around to cutting it. Doing the same job using the cut-as-you-go method would take twice as long.

These two tricks can make the difference between earning $150 a day or $250: First, always lap back your cuts. Second, leave the saw in the truck until you really need it.

■ Sheathing with Exposed Boards

Some ceilings get sheathed with 2 x 8 tongue-and-groove pine that will be exposed. They usually have large beams that are also exposed, as in Figure 15-7. Always put the best side down when laying material that will be exposed. But if only one side has a "V" groove milled into it, of course you'll have to put that side down. The quickest way to do this job is to have a stack of wood laid out directly in front of you (Figure 15-8). Always start with the

Figure 15-8   With this type of sheathing, each piece needs to be cut to fit. Have a stack of wood directly in front of you so that you can just reach up and grab one when you need it. If you come across pieces that are overly crowned, use toenails to drive them together. Try sending two or three toenails in at once on the real tough ones.

groove facing you on the board you're reaching for, and the tongue facing outward on the board you just nailed. Drive the groove of the new board onto the tongue, and then use toenails to snug the fitting together tightly.

In the Southwest they're very fond of using exposed wood in ceilings. Most exposed "vigas" (round logs) are covered with a ceiling of aspen or cedar "lattillas." Lattillas are round or halved sticks that are from 2 to 6 inches in diameter. Figure 15-9 shows a ceiling with exposed vigas which have been sheathed with halved cedar lattillas that were run with a 45-degree herringbone pattern. This can be a very labor-intensive process. The lattillas are far from perfect along their edges and each piece needs to be shaved a little to fit along the imperfections of its neighbor. The labor is worth it though. This is a very fine ceiling, far from the typical drywall you see in most houses.

In the old days (and even today on some reservations) 3 feet of dirt was laid on top of the lattillas for waterproofing and insulation. When doing remodels on older houses in New Mexico, I've often run across this layer of dirt.

■ Nailing the Sheathing

On very steep pitched roofs (6 in 12 and up), nail down cleats so you can walk up and down the face of the sheathed roof. Use scrap 2 x 4s that are about 2 to 3 feet long for this. When you get a sheet nailed down, nail a couple of cleats on either side of its top edge. Then you can step up on them when you lay the next row of sheets. Leave the cleats on the roof for the nailer and the roofers to use. The roofers will peel them off as they go up.

Figure 15-9 In the Southwest they like to use "vigas" or round logs as exposed ceilings. These typically get sheathed with cedar or aspen "lattillas." This old technique has been "discovered" by the modern designers and can be found on many big custom homes in the Southwest.

Nailing off a roof is basically the same as nailing off a floor. Find the nailing schedule in the plans or from the supervisor and nail away. The nailing schedule will sound something like 6 and 12. That means they want a nail every 6 inches along any supported edge, and a nail every 12 inches in the field.

Remember, when nailing the starter board, always angle your nails on the outlookers. Be careful when nailing the edge of the starter board along the fascia and frieze block. They're usually very brittle and tend to split under the nail gun. A lot of framers just nail down into the rafter tails and forget trying to nail the edges. This holds the starter board and

avoids tearing the material to shreds with the gun. Since the starter board is beyond the exterior wall and has no real structural value, the nailing schedule doesn't apply.

The Pay

Roof sheathing and nailing can pay from 5 to 15 cents a square foot of sheathing applied. You don't go by floor square footage, but roof square footage. There are a number of factors that affect the piece price. Here are some of the variables:

- Are there starter boards?

- Is it all straight runs with no valleys?

- What's the pitch of the roof?

- Is nailing included? Sometimes a roof will pay 5 cents a foot to sheath and then another 5 cents to nail. In some cases the same crew doesn't do it all.

- Are nails supplied?

- Are you doing the California fills for the roof stackers?

- Is the economy keeping everybody busy, or is the work going to the lowest bidder regardless of quality?

The people who make the most money at sheathing have invested in a large trailer-mounted compressor and have hired a few helpers to lay down sheets. The helpers will sheath the houses and then the boss comes behind them and nails. Most of the work is in the actual sheathing. You should be able to do the nailing and keep up with two or three crews doing the sheathing. That's assuming the work is out there to keep those crews busy.

On a tract, two people can usually sheath a house in one day. With two crews you'd have two a day to nail. Let's say you had 3,000 square feet of sheathing per building. Two crews would give you 6,000 feet a day to nail. If you split the 10 cents for each house three ways with each crew, you could average $200 for about five hours work per day. I can live with that.

Exterior Elevations

Back in the old days, when a developer decided to build a housing tract, the process was pretty simple. The developer and the architect would argue over possible styles (in most cases the "ranch house" won), decide on a roof line, maybe toss in some siding, and away they went.

Today, teams of designers lecture architects, who fight to stay one step ahead of each other and the planning department — not to mention the neighborhood planning commissions! The developer hopes his designers have added the right decorative touches to woo homebuyers into thinking they're getting more than just wood and stucco. But that's not all bad news. In fact, it's good news for us in the labor pool. All the exterior frills not only line the developer's pockets, but for once they stand to line yours as well. More frills mean more work.

When designers want to dress up the exterior of a building, they'll add arches and build-ups to break up the monotony of an otherwise straight wall or column. These forms can accentuate a column with a stronger, massive look, or simply add an interesting highlight that will create shadow and depth.

A *build-up* is layers of wood, possibly ranging from a 2 x 10 at the bottom to a 2 x 4 on top, placed on columns, at the base of an arch as shown in Figure 16-1, or along window sills. A *band* is similar to a build-up except it runs along the length of the wall at a set height. Take a look at the bands in Figure 16-2. These might simply be one 2 x 4 wide, or wider, depending on the design.

If you want to see a picture of what the house will look like when it's complete, turn to the *elevations* page on your set of plans. In framer's jargon, any exterior stucco build-up is usually called an *elevation*. A good

Figure 16-1   Elevation bands of layered material are usually applied at the point where an arch ends into a column and then at the base of the column. These arches are a good example of flattened-out arches.

set of plans will specifically call out the radius size of each arch. Some (especially on tract projects) will even call out the size of wood used on each layer of build-up or set of bands. On a not-so-good set of plans, you'll have to scale it off the drawings or make it up as you go.

Laying Out Arches

The most difficult part of building most elevations is laying out the arches. There are two kinds of arches. First, the *true arch*, which is a radius true to the opening. Figure 16-3 shows an example. Second, there's the flattened-out (or *lazy*) arch, with a radius that's larger than a true radius. See Figure

16-4. But what's a true radius? It's a radius that's half of the opening size. If the arch fits in a 10-foot opening, a true arch has a radius of 5 feet.

As an arch flattens out, the end points at the sides of the arch raise up. The designer might do this for practical reasons (so you won't run your car or your head into the sides). Or it might just suit the design of the house better.

■ True Arches

We'll start with a true arch, using the arch in Figure 16-5 as an example. The opening between the two columns is 91 inches. Divide that in half to get a radius of 45½ inches. Now you're ready to lay out the arch. First, clean out a large area in the house

Figure 16-2   Here are some typical 2 x 4 bands. The stucco will span from one band to the next, so this will be one 3-foot band, not four individual bands.

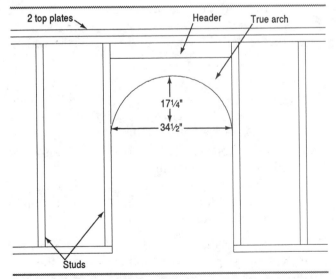

Figure 16-3   A true arch is one with a diameter that matches the exact width of the opening it was designed for. If the opening is 34½ inches wide, the radius that you'd use to build the arch would be 17¼ inches.

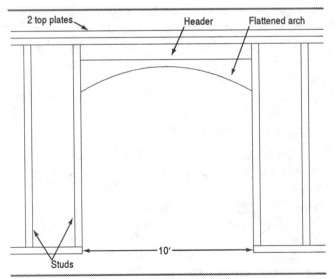

Figure 16-4   A flattened-out arch will have a diameter much larger than the opening itself. This raises the sides of the arch. In this drawing you can see that we used only 10 feet out of a radius that measured 64 feet.

Figure 16-5 This is an example of a true radius arch. At this point we're hanging and nailing off the arch to the columns. Once the sides and top are in place, nail off all the edges.

to work in. In the middle of the room, snap a straight line with your chalk box. Now find a piece of plywood that's the size of the opening plus any overlap you'll need on the top and the sides (if the arch mounts outside of the opening). Place the plywood toward the top of the line, making sure the center of the plywood is exactly over the line that represents the middle of the opening.

To ensure that the plywood is centered on the line, I like to snap a line down the center of the plywood sheet. For an 8-foot piece of plywood, you'd snap a line at 4 feet. Then line up the top and bottom points of the snapped line on the plywood with the snapped line on the floor. Then you know the plywood is perfectly square and lined up with the line on the floor.

The top measurement. Now you'll need to figure out how much wood to leave at the top and sides of the arch so you can attach it. On the arch shown

above, we needed 18 inches above the top of the arch (the distance that the elevation plan showed from the center of the arch to the top of the wall).

On some jobs, this can get tricky. In Figure 16-5, there's a 12-inch-wide column on either side of the arch, with a beam supporting the roof set along the front of the two columns. In the back, the columns rose to the height of the ceiling. So we had two different top heights to consider, one for the front piece of the arch and another for the back. Since an arch is constructed of two pieces of plywood joined by blocks, these pieces needed matching radius cuts, but varying top and side cuts.

We laid out the back side of the arch with the full 18 inches of plywood from the center of the arch to the top of the plywood. On the front piece, the arch mounted up to the 4 x 10. There was also a 2 x 4 top plate on top of the beam. This totaled 11 inches, which we subtracted from the 18 inches. So

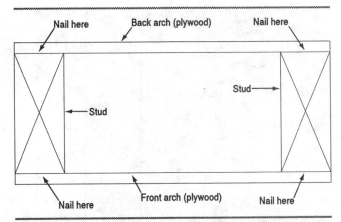

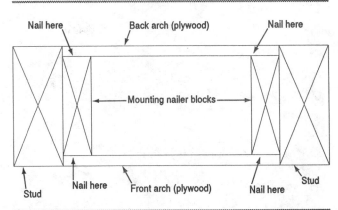

Figure 16-6 One way to mount an arch is to lap the sides over the framing. In this plan view, you can clearly see the overlap. For this method, you have to calculate the side measurements when laying out the arch.

Figure 16-7 The other way to mount an arch is called flush mounting: the outside of the plywood ends up flush with the framing it's mounted against. For flush mounted arches you need nailer blocks on the supporting framing lumber between the front and back arch pieces so you have something to nail the arch to. These nailers will have to be ripped down to fit. For instance, if you have 2 x 4 supporting lumber and two ⅜-inch plywood arches, you'd rip down the nailers to 2¾ inches (3½ - ¾ = 2¾).

the front arch had only 7 inches from the center of the arch to the top of the plywood sheet. The front and the back pieces needed different top heights to end up with the center of the arch at the same height from the ground.

The side measurement. Now we need to figure the side widths. An arch can be mounted two ways. Either lap the sides of the plywood over the columns, or cut the plywood to fit exactly between the columns. In Figure 16-5, I decided to lap the front piece over the columns because the whole column and roof structure was a little top-heavy. By lapping the front of the arch over the columns, I added some shear strength which made the whole structure more solid. On the back, I cut the arch to fit flush between the columns.

Since I lapped the front and mounted the back flush on this arch, I had two different top heights and two different side widths — yet the radius end up matching perfectly. I've gone through this in detail because it's as complicated as a basic arch will ever get! If you follow this process, you can figure and cut any arch.

Figure 16-6 shows a plan view of an arch that's lapped over the columns. Figure 16-7 is an arch that's mounted flush on both sides.

Now let's return to the plywood laid out on the reference line we snapped. With the top and sides established, we can go ahead and mark out the front arch. Pull down 7 inches from the top center of the wood along the snapped center line, and make a mark. This is the top center of the arch. From that 7-inch mark, pull down another 45½ inches (half the width of the opening) and make a mark on the snapped line on the floor. This is the point you'll use to pull the radius.

Set a Teco nail into the slab at this point. Gently hook the nail with the end of your tape and pull it up to the 7-inch mark. Holding the pencil on your tape at the 45½-inch point, mark out the radius on the plywood, being careful not to pull out the nail (Figure 16-8). Note in the photo that we had to add pieces to the plywood to make the sides extend down far enough (52½ inches in our example: 7 + 45½ = 52½). Figure 16-9 shows all the dimensions clearly as well as the arch line drawn in.

To summarize, the finished front piece has the sides lapping 12 inches over the side columns, and the top ends 7 inches under the header. The back piece will have the same radius, but with different

Figure 16-8 Pull the radius as smoothly as possible so the line comes out true. Hook the tape on a nail set on the line that runs directly through the middle of the arch.

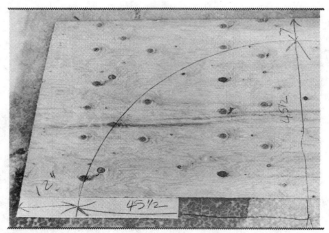

Figure 16-9 Here are the calculations that let us pull the front lapping arch: 45½ inches is half the opening, 12 inches on the side because we're lapping instead of mounting flush, and 7 inches on the top. For the back flush arch we eliminate the 12-inch side cuts, and extend the top to 18 inches.

top and side measurements. Because there's no header on the back, the top will be a full 18 inches. The sides, because they're mounted flush, will be zero.

Cutting and blocking. Now cut the line you made. On round cuts, it's usually easiest to cut halfway through the first time, then do another pass at full depth. We're using ⅜-inch plywood, so it cuts very easily (Figure 16-10).

After you've cut the front and back plywood, you'll need to cut the supporting blocks. The columns measure 12 inches wide. Since the back side is mounted flush, we'll cut the blocks ⅜ inch less than the column width to allow for the thickness of the plywood. That makes 11⅝-inch blocks. Place the front and back sides of the arch together and then lay out the blocks, leaving a few inches between them.

First, make sure the front and back plywood arch curves match perfectly, regardless of the top and side measurements. Then use the 3½-inch face of a small piece of 2 x 4 to make your block layout marks. Hold it along the cut edge of the arch and mark each side of the block so you're making a 3½-inch layout on each piece of plywood. See Figure 16-11. If you have a flush-mounted arch, keep the side blocks from encroaching on the mounting nailer blocks by moving them up a little from the bottom of the arch. Figures 16-7 and 16-11 show the location of the mounting nailer blocks for a flush-mounted arch. Lay out the blocks with a couple of inches between each block to give the stucco crew enough space to attach their paper and wire.

Once you're laid out, count the number of blocks you need to cut. Our blocks were cut to 11⅝ inches. Now nail up the blocks to each layout mark on the front face plywood of the arch, keeping the blocks flush with the cut edge. Get two nails through the plywood and into your blocks. A nail gun works great for this task.

Once the front face is nailed, flip the arch over and nail the back face to the blocks, being careful to keep the front and back layouts in line. See Figure

Figure 16-10   Cut out the center of the arch and discard it.

16-12. If you laid out the front and back together before you started assembling the arch, the arch should frame up square regardless of the difference in top and side dimensions.

Preparing the columns. Now that the arch is built, you can detail the columns to accept it. If the arch laps over the front of the columns, there's nothing to do there. If the arch mounts flush on the back side, you need nailer backing on the inside of each column and along the ceiling joists to nail the arch to. A piece of 1 x 4 works well, or you can double up some scrap plywood strips. Nail the nailer backing directly to the column, buried between the front and back arch pieces.

Use a piece of 3/8-inch plywood (the thickness of the arch material) as a gauge to mark the location of the nailer. Keep it back from the outside edge of the column by this distance. Nail the nailer to the sides

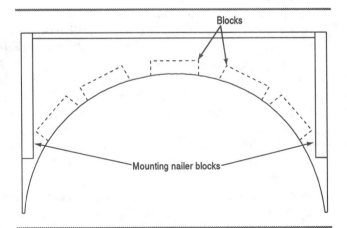

Figure 16-11   Lay out the middle blocks for the arch using a scrap piece of 2 x 4. Space them about 2 to 3 inches apart. Notice the side mounting nailers. You want to keep the first block up a little so it doesn't interfere with the mounting nailers. If you put them tight to the sides, you won't discover the problem until you go to raise the arch.

Figure 16-12   Nailing the blocks to the plywood arches is best done with a nail gun. Nail up one side and then flip it over and nail the other side. Since we laid out the arches together, these layout marks should keep the arch square when we nail it together at this point.

Figure 16-13   Building a scaffolding is easier than working off ladders. Slide the arch into place, starting below the side nailer mounts and then up into place. Once the edges of the plywood are all square, go ahead and nail it up.

and the top, running full length along the top. Along the sides, run it down the length of the side measurement of the arch (45½ inches in our example).

Mounting the arch. It's usually worth the time to build a quick scaffold when you're ready to mount the arch. (You can see the scaffold in Figure 16-13). You may need someone to help you slide the assembled arch up into place. Check that it's sitting square and tight in the flush-mounted area along the back sides and top, then nail it up. If you have a nail gun, place it within reach before raising the arch. On this job, we had the gun hanging from a nail tacked to the column. With the arch set, we could easily grab it and shoot the arch.

Door or window arches. Sometimes you'll have interior door or window openings with arches over them. When building the walls, I like to frame these openings with the headers high to leave plenty of room to install the arch. If there are a lot of arches to cut, I recommend setting up one carpenter with enough space to organize these arches. If you aren't organized, you'll end up wasting hours of time and stacks of wood.

Depending on wall heights and door widths, you may end up with a variety of different arch sizes. Make sure that each arch has a code and a label telling whether the plywood is the back or face.

An arch for a door opening is usually a true arch, mounted flush with the wall framing. The true arch being installed in Figure 16-14 is a good example. The door had a 34½-inch opening, so the arch was cut with a 17¼-inch radius. The builder gave us the rough opening dimension from the floor to the top center of the arch. This established the distance from the top center of the arch to the header. Notice the label on the right side of the arch in Figure 16-14 that reads 5 over 34½. There's 5 inches of plywood above the arch, and an opening of 34½ inches. That's how this framer organized his arches.

On thin arches for doorways or windows (3½ to 5½ inches wide), it's important to keep the front and back pieces of the arch square with each other when you nail them together. A large arch will be really flexible once it's built. You'll be able to move it around in your opening when mounting it to square up the edges. But a small doorway or window arch

Figure 16-14   Interior arches will almost always be built and mounted flush, so there's no interference with the drywall. Notice how the header was framed in high. This allows room for the arch blocks to go full circle. Also notice the notation on the arch. These numbers mean a 5-inch top with a 34½-inch width.

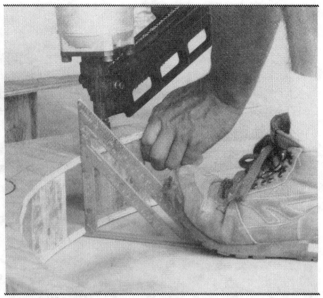

Figure 16-15   On thin arches such as this one, it's imperative that they're built square. If they get just a little off, they'll be a bear to get in place. Larger arches will flex when mounting. The smaller the arch, the tighter it is.

will be solid once it's built. When you're building small arches, use your speed square set along the floor to keep the front and back plywood arches square when you nail them together (Figure 16-15).

Barrel arches. On large customs or commercial jobs, you may have to construct arches that are too large to build on the ground and lift in place. These *barrel arches*, as they're commonly called, are usually stick framed.

The diameter of the arch in Figure 16-16 was nearly 20 feet, which would make it impossible to raise by hand. Not only was it wide, it was 10 feet deep. If you look closely, you'll see that the top of the door opening had an arch built to accept the

radius door unit. There was also an arch that surrounded the rough door opening to give the stucco some detail around the opening. The barrel arch was built 6 inches wider than the detail arch. Once we knew the radius of the detail arch, we could simply add 6 inches and come up with the radius of the barrel arch.

The first step in building a barrel arch is to build some scaffolding to work from. On this arch, we chose to use an intermediary plywood arch located in the center and nailed to a ceiling joist to help carry the load of plaster that the arch would have to support. So we cut three plywood arches: one for the outer edge, one for the center, and one that rested against the wall. Of course, they all had different top and side cuts. Then we put them together and made layout marks on all three at once.

Finally, we installed the outer and the middle pieces. The middle arch required two side studs for support. Then we cut 5-foot-long blocks to span from the outside arch to the middle arch. Using a nail gun, we hung the blocks from the outer plywood to

Figure 16-16   Here's an opening ready for a barrel arch. Notice the radius doorway had another stucco detail radius framed 6 inches wider than it is. The barrel radius was another 6 inches wider than the stucco detail, so we had a radius to work with.

Figure 16-17   After the front and middle plywood arches were hung, we built half the barrel. We built from the front plywood arch to the center first.

Figure 16-18   The middle arch was nailed up to a ceiling joist with a serious amount of nails. The second half of the arch, from the middle to the wall, was stick framed like the first half. Later, we ran legs up from the arch blocks to the ceiling joists for extra support.

the inner, as shown in Figure 16-17. The last step was to hang the plywood arch along the wall. We took a measurement for the blocks that spanned from the middle arch to the wall arch and placed them individually, as shown in Figure 16-18.

■ Lazy Arches

To build a flattened-out (lazy) arch, begin with the width of the opening. Then figure how far you want the curve to come down on the sides. You can usually scale the sides off the plans. Cut the plywood to the total width of the opening that the arch covers, and snap a line along its center. Next, lay out the plywood as you did for the true arch, centering it on a snapped line. Finally, pull down the measurements on the top and sides of the plywood.

The garage door pictured in Figure 16-19 is an extreme example of a flattened arch. The garage door rough opening left us with only 1½ inches on the top. For design reasons, the builder decided to make the arch 6 inches wide on the sides. We pulled down 1½ inches on the center of the arch and made a mark. Then we made a mark 6 inches down on both sides. We want the radius to intersect these three marks — but how do you pull a radius dimension out of thin air to connect these three points in a perfect arch? I've seen people just mark out an arch by eye. But I don't do it, and I don't recommend it for you either!

Once you know the width of the opening and height of the arch, there are two ways to find the radius. The first is to work it out with a tape. The second is with a calculator. Let's do it the hard way first, which is with a tape.

Figure 16-19   Here's an extreme example of a flattened arch. This is the way we pulled the arch that's shown in this garage door. At 64 feet, it's easily the biggest arch I've ever pulled! Using a string and a 100-foot tape was essential. When it was stuccoed, it looked great.

Finding the radius with a tape. With one person holding the tape on the snapped line and the other checking against the top and side marks on the plywood, you can hunt down the radius. You need to find a number on your tape that matches at the top center mark, as well as the side marks.

Sometimes you'll need to extend a string line in line with your snapped line 60 feet or so in order to have room to work. See Figure 16-19. With some-one holding the tape end on the string, you'll move it from the top center mark and down to the side marks to see if it matches. For instance, you might start out at 40 feet at the top center 1½-inch mark. When you move it down to one 6-inch side mark, the tape reads 34 feet. The arch isn't working. The numbers have to match. Have your partner move down the string and try 50 feet at the top mark. Maybe you'll get 48 feet on the side mark. Your

numbers are getting closer, so at least you know you're moving in the right direction. Keep moving the tape up or down along the snapped line or string until the number you get at the top of the arch matches the sides when you swing the tape.

The important thing here is that the tape must pivot from the center line, like scribing a true arch. Sometimes you'll move the tape 3 feet and it only changes the reading ½ inch. Just keep trying. In Figure 16-19, it was a 64-foot radius! Just don't give up. The radius is in there somewhere.

Finding the radius with a calculator. For you calculator buffs, there's an easier way to figure a flat arch. Here's the equation:

$$R = \frac{B^2 + A^2}{2A}$$

where:

R = the radius

A = the height of the arch from its highest point to where it dies into the sides

B = ½ the width of the arch or opening

In our example, A = 6 inches and B = 96 inches (since the opening is 192 inches):

$$R = \frac{B^2 + A^2}{2A}$$

$$R = \frac{96^2 + 6^2}{2 \times 6}$$

$$R = \frac{9216 + 36}{12}$$

R = 771 inches

So the radius is 771 inches, or 64.25 feet.

Obviously this is a much quicker and more accurate method. Do yourself a favor. Learn to do these equations with your calculator — and always carry your calculator (and the equations) with you. They're like any other tool. You can't use them if you don't have them.

Figure 16-20 So you want to piece out some arches, do you? Study this for a while, and learn. It really makes a lot of sense after a while! Of course the calculations provided by the architect helped matters out considerably.

Figure 16-21 Here's a little fluff surrounding a garage door opening that I imagine turned into a real head-smacker. This is a pretty unusual piece, as the gable truss is supported by a couple of cantilevered 2 x 6 blocks! I suppose once the rounded elevations are stuccoed, they'll also give some support.

Complex arches that plane in to one another (as shown in Figure 16-20) can be very time-consuming. I'd be wary of taking on a job like this on a piecework basis. Finding the points of intersection was the main problem. The architect on the job helped out by sharing the sheets of calculations he used to find the differing arch radii for this beautiful portico entryway.

The Pay

When doing elevations, you'll run into quite a variety of jobs. Exterior pot shelves, window boxes and molded designs for plaster (as in Figure 16-21) can be good money-makers. At one tract I had to build 80 window boxes in two different styles. After I built the first two, I had a list of all the measurements I needed. I paid a laborer $5.00 apiece to cut up the packages, and stored them all in one house. With the nail gun, I could build and install one an hour — at $45 each. Because the builder learned that I was organized and could keep up with his schedule, I spent another two months at that job building arches and applying stucco build-ups around all the windows.

The piece price for elevations can vary from $50 to $500 per house. To make good money, a nail gun is a must. If I'm trying out a job to see if it pays enough, I'll do one house without the gun. If I make a respectable hourly wage, I know their price is good. If you do the first one with the gun and you aren't making good money, you know something is really wrong. It's time to renegotiate, or walk on.

Drop Ceilings

O nce the roof is stacked, sheathed, and nailed off, you can remove the interior plumb and line braces. Then it's safe to start work on the interior of the building. A lot of the interior work consists of fixing shoddy work, making changes, and general work that couldn't be included in any of the piecers' duties.

On a tract, most interior work is handled by pick-up or hourly carpenters. But there's an exception to that rule. Ceiling work, if there's enough of it, is usually pieced out. Pick-up carpenters often don't want to do the drops, or aren't experienced enough to do them. If you show up at the right time, with the right tools, you can usually piece out the ceiling work. By the way, the "right tools" in this case are a nail gun and compressor. The contractor will usually supply the nails (which are expensive) or compensate you for the nails you buy and use. It's not impossible to nail together countless small pieces of wood overhead by hand, but it can be very frustrating. With a nail gun it's almost fun.

Depending on the design, some houses have ceilings that need a considerable amount of attention after the roof is stacked. Some ceilings need to be lowered (or *dropped*), some need to be vaulted. Too often, the house designer doesn't work out adequate access for the plumbers and heating contractors. Of course, they've got to provide a sketch to let the planning department know that the proper size and amount of pipes will be installed. Unfortunately, on some houses the schematics turned in to the planning department are more wish than reality.

Split-level houses and houses with flat roofs or low-pitched scissor trusses can be a nightmare for the heating subcontractor who's trying to run the ducts. If the house has no available space to use, the framers

Figure 17-1   This house had a shallow pitched roof (4/12) with scissor trusses. The architect's dream was to run the ducts in the trusses. But in the real world, the heating sub had to run all the ducting under the second-story floor joists and then we buried it with a flat-joisted drop. There was simply no room in the trusses. Luckily, the lower floor had 9-foot walls or this would have been more of a disaster than it was. Side walls were built and attached to the floor joists first. Then we toenailed the flat joists with a nail gun to the bottom plates of the walls.

probably have to install false ceilings in hallways or closets to create space between the joists and the new ceiling. These *dropped ceilings* hide the ducting as it travels from one side of the house to the other (Figure 17-1).

Not all drops are built to hide other subcontractors' equipment. Many kitchens have elaborate drops incorporating pot shelves and hidden recessed lighting. This may vary from the standard fluorescent box to a drop that incorporates both indirect lighting and plant shelves. In custom homes, you might run across dining rooms with an elaborate ceiling as a backdrop for the chandelier (Figure 17-2). Rooms with high ceilings probably have the closet ceilings dropped down to 8 feet to keep them from looking like elevator shafts.

Custom designers often use tall walls so they have space to dress up their ceilings. Even a small band of indirect lighting around the perimeter of a room needs at least an extra foot of ceiling height. The more elaborate sculpted indirect drops (Figure 17-3) need a good 2 feet to be effective. This puts the walls at a minimum of 11 feet.

Building Basic Drops

Most drops are built simply to lower the ceiling. With this in mind I'll go through the steps to build a basic square closet drop. The principles also apply to many other drop configurations. For this exam-

Figure 17-2 Drops can vary from the easy (in Figure 17-1) to the elaborate, as shown here. This layered drop was designed to have a chandelier hanging from its center. We have everything here from flats to curves to circles, and then some.

ple, I'll use a closet that's 5 feet by 10 feet. Perhaps the main ceiling is 14 feet high but the dropped ceiling in the closet will be a more reasonable 8 feet.

To begin with, snap a line 8 feet off the floor on each 10-foot-long wall. Then snap another 8-foot-high line along each 5-foot-long wall. Cut two boards to fit along the 10-foot walls to make the *rims*. For a 5-foot span, 2 x 4s on edge will be strong enough for the rims and the joists.

Before laying out the joists, consult the supervisor to see if any special layout is needed. Sometimes the plans call for a light centered in the closet. If you put a joist in the center, it'll have to be changed later. Sometimes an attic access door is installed in the closet ceiling and you'll need to head out a 24 x 32-inch opening for it. After you're clear on the lighting and any openings, match up the two rims on the ground. Then pull a 16-inch on center layout for the joists on both rims at the same time.

Figure 17-3 Here is a sculpted drop with indirect lighting that has been drywalled. The bulbs are mounted on the top of the drop. They'll reflect off the ceiling and create a diffuse glowing light. Large sculpted drops need tall walls to be effective and not seem overwhelming.

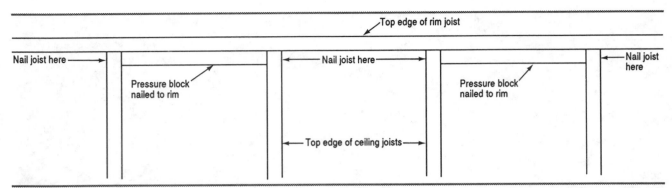

Figure 17-4 For most drops you can substitute pressure blocks for hangers. Pressure blocks go in every other bay and are nailed directly to the rim. Then, when the joist is raised in place, it's face nailed into the block and toenailed into the rim.

Now, find out from the super if you have to use hangers, or if pressure blocks are acceptable. If you'll be using hangers, go ahead and nail them to the rims on each layout mark before you put up the rims.

On most jobs you can use pressure blocks instead of hangers on drop ceilings. If you can use them, put them on the rim every other bay between the joists, as shown in Figure 17-4. For a 16-inch on center layout, they would be 14½ inches long. Nail them directly to the rim and then face nail the drop joists to the pressure blocks and toenail them to the rim. Whether you're using hangers or pressure blocks, nail them on before you go up with the rims.

To install the rims, set a nail in a stud a little below the snapped line on one side of the 10-foot wall. This will hold up one side of the rim as you nail the other. Holding the bottom of the rim along the snapped line, nail one side of the rim with two nails into the corner stud. Now make your way from one stud to the next, adjusting the rim up and down as you go, to meet the line. Once you have both rims up, check the measurement for the drop joists. Check each side and the middle, as they might not be the same. For our example, the joists would be 5 feet minus 3⅛ inches to allow for the rims, or 56⅞ inches.

If you're doing a drop for a long hall and the middle joist measurements change dramatically from one end to the other, check to see which wall is bowed. If you can fix it easily, go ahead and do it. If it looks pretty messy, I suggest bringing it to the supervisor's attention. Let him make the call

whether to leave it or not. You have to walk a fine line between bugging the super too much or making the judgments yourself. After one such question, you'll get a feel for his standards so you'll know when to ask in the future. Some want it perfect; some don't even want to know about it. Just remember that if something's way off and you didn't warn anybody before you built the drop, you'll probably be held responsible later.

Once you've figured the size of the joists, cut them and crown them. Now install a hang nail on one side of each joist by putting a nail in the top and bending it over. Working on the opposite side of the hang nail, slide the board in place. It should fit snugly. You don't want to have it too oversized or it'll push the walls apart. If it's too undersized you'll have a hard time hiding the gaps.

Nail up one side of the joist flush to the bottom of the rim. Then nail up the other side to the opposite hanger or pressure block. On the two end joists, follow the line you snapped on the wall. Check along the lower edge of the rim to make sure the fire blocks in the wall aren't leaving any gaps. In fact, make sure the fire blocks are there at all. They're usually put in when the walls are built. If not, cut and install them in each bay so they're half above and half below the rim. Installing fire blocks should pay extra on a tract.

Although some drops get really elaborate, they're never anything more than rims, joists, and pressure blocks or hangers. And many times you'll

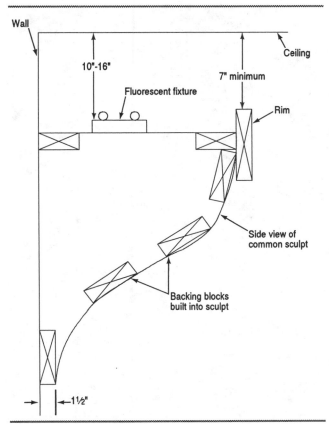

Figure 17-5 Here's a cross section of a sculpted drop. You should allow 10 to 16 inches between the ceiling and the light fixture to avoid any hot spots. Leave at least 7 inches between the rim and the ceiling for the light to escape.

be designing them as you go. There are rarely blueprints for drops. You'll get an elevation showing what it should look like, or you might get a tour through a finished model on a tract. Then it's up to you to figure out how to make it look like the model.

Building Custom Drops

Some custom homes will have a plan for the more elaborate drops. Always try to consult with the electrician and the heating and air people before you start. Find out any special bays they need before you hang the drops. You'll save a lot of notching

later on. You'll often build the drop to oblige them, so in effect they'll be giving you the drop's specifications.

On the upper end of the custom home market, you'll find designer touches intended to provide a personal statement by the designer and the owners. Unfortunately, important issues such as energy efficiency often take a back seat to grandeur and square footage.

Indirect lighting is one of the favorite places for designers to add their special touch. Instead of having bare bulbs lighting the house, the drop hugs the wall around the entire room and hides the lighting system. Fluorescent bulbs aimed upwards bounce light off the walls and ceiling, creating a pleasing diffused light. These drops can run down halls, above kitchen cabinets, and even in layers around bedrooms. Because they're built with wood and then drywalled or plastered, the shapes and sizes are limited only by the ceiling height. The bulbs should be a good 10 to 16 inches from the ceiling. That allows the light to disperse without creating hot spots.

■ Sculpted Indirect Drops

Sculpted or curved drops are usually only seen in the high-end custom or commercial buildings. But there's no reason, except possibly financial, that these drops can't add some personality to a smaller custom. They're quite easy to build. I'll go over the steps in building a sculpted drop. You can use this same method to build a recessed lighting drop that uses a more conventional square design, which is much easier on your drywallers.

Designing the sculpted drop. The first step to building a sculpted drop is to decide on its shape. The most common design employs a slender bottom that grows into the curve at the top, like the one in Figure 17-5. The base should leave the wall at an abrupt 45-degree angle rather than flowing into the wall. This will give the drop a shadow line which helps to separate it from the wall and gives the sculpt a distinct starting point. The top of the sculpt should have about a 4-inch plumb cut on it. This

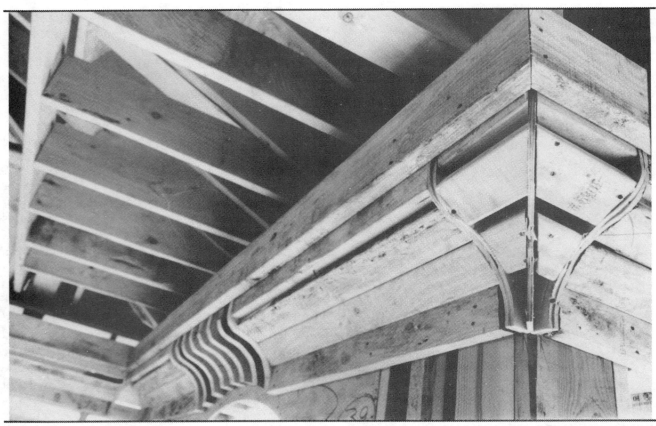

Figure 17-6 Outside corners require you to fabricate hips. In this case, the top of an arched door ran into the bottom of the drop. If we raised the drop, we wouldn't have enough space between the rim and the ceiling joists to allow sufficient light to get out. We solved the problem by cutting common sculpts out of 2 x 10s and following the radius of the door at a constant 2 inches where the drop ran into the door.

allows for a 2 x 6 rim, held above the top of the sculpt 2 inches, to be mounted all the way around the sculpt (Figure 17-6). This rim gives the sculpt a straight, finished line around the top, and also helps to hide the lighting. Make sure you leave a minimum of 7 inches between the ceiling joists and the 2 x 6 rim to allow the light to reflect out into the room.

It may take a few practice drawings to get a design that you really like. Sketch out a few and then pick the best from your selection. Once you have a design, use a jig saw to cut the pattern out of a piece of plywood. This will be your *common template*. We'll use a lot of roof terms for sculpted drops (commons, valleys, hips, jacks), so don't get confused by this.

Building the sculpted drop. A sculpted drop is put together in 6- to 8-foot sections. We'll run a combination of 6- to 8-foot 2 x 4s and 2 x 6s flat between a set of common templates (Figure 17-6). This ensures that the drywaller has enough backing to make the radius work. And this is a great place to use up all that scrap wood lying around the job.

When the sculpt approaches an inside corner, it makes the turn through what I call a valley (Figure 17-7). As it approaches an outside corner, it makes the turn through a hip (Figure 17-6). These hips and valleys are extended and compressed versions of the common template. You can find their shapes by using strings pulled over two commons into a piece of plywood set up appropriately for a hip or a valley.

Figure 17-7 This is a typical 30-degree valley built and ready to install (shown upside down). The two side pieces of plywood are commons and the middle is a valley. Notice how the jack blocks are labeled with their lengths and angle cuts. There are also two blocks on the bottom back side.

Figure 17-8 You can find the shape of a valley using a jig, a string and a couple of pairs of hands. On the piece of plywood, place a dot where the string hits it at ten or so positions. Then connect the dots, cut the valley, and check it in the jig to see if it's shaped right. If you have angles other than 45 degrees, set up the jig for each differently-angled hip or valley. You can also use the jig to figure the size of the jacks.

Notice the temporary setup in Figure 17-8. A common is held up in the left foreground and a string is pulled over its top edge and toward the valley setup in the background. The back edges of the commons must be in line, and the valley setup should be held square with temporary blocks as shown. The valley setup consists of two commons at 90 degrees to each other and an uncut piece of plywood set between them at 45 degrees.

Looking at the common template in the foreground of Figure 17-8, notice the lines drawn every inch or so along its edge. The valley common has identical lines. The person in the foreground and the person in the background hold the string on the same line. This will send the shape into the uncut valley plywood if the background person runs the string past the valley common and into the uncut valley plywood.

Notice how the background person in **Figure 17-8** is making a small pencil mark on the uncut valley plywood. As these two carpenters move from one line to the next along the common templates, they're making marks on the valley where the string hits. Once all the lines on the commons have been transferred into dots on the valley, you can pull the valley board and then play "connect the dots." After all the dots are connected you should have the shape of the valley drawn on the plywood.

Figure 17-9   Having a large work area with a temporary table set up really helps. The hips and valleys should be assembled using staples or 8d nails through a nail gun. Trying to do this by hand can be very frustrating!

Figure 17-10   Assemble in groups all hip/valleys needed for each room. These four 15-degree and four 30-degree valleys were installed in the room shown in Figure 17-3.

Go ahead and cut it out and then immediately label this as a valley. Each valley, hip and common will have a unique shape, so it's important not to mix them up. You can establish the shape of the valleys and hips easily using this string method. Once you have one valley and one hip cut, you can use them as patterns to mark and cut as many as you'll need. The shapes won't change as long as you use the same commons.

But here's something to keep in mind: If your room has walls that aren't set at 90 degrees to each other, you'll have to figure out the valley or hip for each angle. The valley shown in Figure 17-7 was for a 30-degree inside corner.

The best way to do it. After doing a number of houses a number of different ways, I've found a three-step method that's the quickest and cleanest way to build a sculpted drop:

▪ First, build the hips and valleys on the ground.

▪ Second, mount them in place.

▪ Third, measure between them for the common sections.

The first time I did a sculpt, we built the corners in place. I thought it had worked pretty well until I pieced out the next house to another carpenter. He put one man to work building all the hips and valleys on the ground, while another built stock 7-foot common sections for the long runs. Needless to say, he made a killing, and I adjusted my price on the next house!

I like to begin by making a general count of how many common, hip, and valley plywood pieces I'll need — then I mark out and cut them up. Keep in mind that a 45-degree valley is slightly different from a 15- or 30-degree valley. You have to consider the angle of each hip or valley. When you've figured a special valley using the string method, label it as *30-degree valley,* and keep it only for tracing. Establishing each hip and valley with the strings takes a while. If you use up your patterns, you'll have to make new ones. And if you don't label them properly, they're useless.

Figure 17-11   The first step to doing any large drop is to construct a scaffolding around the perimeter of the room. Trying to work off ladders will only wear you out.

Once all your plywood pieces are cut up, you need to determine the size and angle cuts of the 2 x 4 jacks that hold the hips and valleys together. Set up two commons and a 45-degree valley in a jig as shown in the top center of Figure 17-8. Now take some measurements for your 2 x 4 jacks.

When you've found their lengths, you'll have to guess at a few angles until they fit properly. Because the jacks are set on a slope, they won't really follow the angle of the corner, but usually will fall a little under. For instance, with the valley pictured in Figure 17-7, the corner was exactly 30 degrees, yet the jacks ended up at 25 degrees. A good quick way to establish the angle is to hold up a short piece of 2 x 4 square with the common, and scribe the angle that the valley runs under it. Make the cut, check its fit, and then adjust it if necessary. The jack that opposes it will have the same angle cut, but cut opposite.

Notice on the 30-degree valley in Figure 17-7 that each jack is labeled by length and degrees. The bottom jack is 3¼ inches to the long point (3¼ LP). The face angle is 25 degrees and the saw's blade is set at 15 degrees. The next jack is 5⅝ inches to the long point, with the same 25-degree face cut and 15-degree blade angle. After you establish these numbers, write them down! Then you can cut the jacks for all the hips and valleys at once.

I like to build a temporary table so that I can comfortably use a nail gun to shoot them all together (Figure 17-9). In Figure 17-10 you can see the finished 15- and 30-degree valleys organized in groups.

The next step is to mount all the hips and valleys. For tall ceilings, set up a temporary walkway around the room like the one in Figure 17-11. Hanging this drop off a ladder is no fun! Next, snap

Figure 17-12 With the line snapped, you can hang the corners. To the left, a 7-foot length of common sculpt was attached off a valley. Then we took a measurement from it to the next valley and built the section that's on the scaffolding, soon to be installed. This bathroom also featured a radius in the drop that matched the round bathtub platform. Notice how the fire blocks installed in the walls during wall framing match up evenly with the bottom of the sculpt.

a line around the entire room that represents the bottom of the sculpt. If you don't have fire blocks in along this line, it's probably easiest to mount them now.

Now hold each hip/valley in place with the bottom on the snapped line and shoot it to the wall. Figure 17-12 shows a few sections up and one waiting on the walkway. If you have long runs between hips or valleys, place 7-foot common sections in from either side. Then take a measurement and build a section for the center. Otherwise, take all your measurements between the hips and valleys and build the common sections for the entire room at once. With a helper, you can lift these sections and shoot them in place when they're on the line.

If you look closely at Figure 17-12, you'll see two lines of blocks in the wall. The lower line was set at 8 feet and was used as the top edge nailer for the shear panel. The block above it was the fire block for the drop. Notice how the bottom of the drop is in line with this row of blocks.

You can also see in Figure 17-12 a radius wall built into the ceiling that wasn't there in Figure 17-11. This bathroom had a radius platform for the bathtub to rest on. The designer wanted the ceiling to match this radius exactly. So we had to hang this radius wall before we could hang the sculpted drop. Its height was determined by measuring from the ceiling, down to the snapped line that represented the bottom of the sculpt. We then had to invent a way to get the sculpt to follow this radius. We solved the problem by cutting and toenailing an individual common sculpt out of 2-by material for each stud on the radius wall, as shown in Figure 17-13.

When you have all the sculpts up, you can go ahead and run the 2 x 6 rim along the top (Figure 17-14). I like to snap a line, 3½ inches down from

Figure 17-13   To achieve the radius in this drop, I first built a curved wall and mounted it up to the ceiling. The wall went from the ceiling joists, down to the line snapped for the bottom of the sculpt. Then I cut common sculpts out of 2 x 10 and nailed them to each stud to create a curved sculpt.

Figure 17-14   Then we installed a drop off the bottom of the wall to flatten out and give a base to the curve. We used three layers of ½-inch plywood to make up the 2 x 6 rim on the curved section.

Figure 17-15   Here's an example of a sculpt held low as an indirect light, and also the same pattern held up tight to the ceiling for an oversized crown molding effect. This is the interior of the clerestory shown in Figure 14-23 back in Chapter 14.

the top of the sculpt, around the entire structure. This ensures that the rim is installed in a perfectly straight line. If the drop is 10 feet or so off the ground, 2 inches worth of rim above the top of the sculpt will be enough to hide the lights.

The drywallers will need to cover the top of the sculpt with ⅝-inch drywall to give the lights a surface to rest on, so make sure the top surface has backing in all its corners for the drywall. They'll either plaster the face of the sculpt or cover it with water-soaked ¼-inch drywall. The water allows the drywall to bend without breaking.

Once all the 2 x 6 rim was up we built a flat drop off the bottom of the radius wall that was built over the tub platform. You can see this in Figure 17-14. The 2 x 6 rim for the radius section of sculpt was made out of three layers of ½-inch plywood.

Sculpted drops can be used with recessed lighting or tucked tight up to the ceiling line to create a very large crown molding effect. Figure 17-15 shows an opening in a ceiling over an entryway that uses both kinds. The sculpt on the top is installed tight to the ceiling and accentuates the 15-degree octagonal angles of the walls. The bottom of the lower sculpt was held a little above the lower ceiling line to give it an extra line. Lighting mounted on its top gave the whole space a soft light in the evening.

The top of the highest ceiling was better than 20 feet up, so we could use the same size sculpt on the top as on the lower one and not risk having it look awkward. If this upper ceiling was closer to the ground we probably would have gone with a smaller sculpt on top. The sculpts tend to look much bigger when tucked up tight to a ceiling than set below them and used with recessed lighting.

There are Styrofoam sculpted drops available that have the same finished appearance as the framed and drywalled units. They're installed, then patched and textured, much the same way drywall is. I've seen sections of these drops, but I've never seen them installed. The builder I worked for didn't like the idea of applying Styrofoam on the wall of an expensive custom home. For one thing, he felt there was a potential liability problem if the paint on the drywall faded to a different value than the paint on the Styrofoam. He also felt they could probably be punctured very easily by an unwary duster.

■ Elliptical Openings

Sometimes, rather than going with a flat ceiling, a designer might create a shallow dip in the ceiling using an ellipse (an egg-shaped or oval opening). If lit well, they almost give the illusion of a giant pillow or cloud hovering in the room. If you look closely at Figure 17-16, you can see the hip and rafters that are covered by this ceiling.

To begin, you'll need to find the size (the length and width) of the ellipse. The plans will probably give you these dimensions. If not, you'll have to figure them yourself. It's best to leave at least 18 to 24 inches around the edges where the ellipse approaches the walls.

Once you have the size, you can frame in the conventional flat drop that goes around the edges of the room. Just remember to leave a rectangular opening in the middle of the room for your ellipse. The opening in Figure 17-16 was framed 1½ inches

Figure 17-16 To construct this ellipse, first we measured and cut the opening. Then we progressively laid out each upper ellipse from its width. Taking it a step further, we could have built an elliptical shelf in line with the lower ceiling, out a foot or so. This would have allowed for the installation of indirect lighting to light up the ellipse, creating a huge diffuse pillow effect. Figures 17-17 through 17-20 give instructions on constructing ellipses.

larger than the ellipse along each edge to allow the framer to cut out the ellipse in a rectangle 3 inches larger than the opening. This is the easiest way to install the ellipse once it's cut out.

Laying out the ellipse. There are many ways to lay out an ellipse, from the very difficult to the fairly easy. I'll go through the *string method*. I think it's the simplest and most effective for laying out large

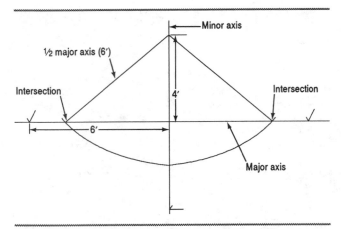

Figure 17-17 The first step to laying out an ellipse is to draw perpendicular lines known as the major axis and minor axis lines. The chicken feet on the major axis mark out the length of the ellipse and the chicken feet on the minor axis mark out the width. Next you pull a radius from the upper minor axis point, and find where it intersects with the major axis. The radius you pull will be half the major axis, or in this case, 6 feet.

ellipses as well as the smaller ones. I like it because you don't need to use any math and the only tool you need is a ball of string.

Begin by laying out the sheet of plywood that will form the elliptical opening in the ceiling. At this point you need to know the length and width of the ellipse. If the plans don't give you any numbers to go by, look at the space you have to work with and estimate a good length. The width should be a little less than three-quarters the length to form a good ellipse. For example, if the ellipse pictured in Figure 17-16 was 12 feet in length, it should be about 8 feet in width.

To describe the process of laying out an ellipse more clearly, I'll refer to the length of the ellipse as the *major axis* and the width as the *minor axis*.

To begin, cut out and lay on the floor the exact length and width of plywood that matches the rough opening you left in the ceiling drop. If the ellipse will finish out at 12 by 8 feet, and you add 1½ inches on each side to make installation easier, you'd need a combination of pieces to get a 147- by 99-inch rectangle of plywood.

Next, snap a line down the middle lengthwise to represent the major axis, and a perpendicular line down the middle across the width to represent the minor axis. These two axis lines can be seen clearly in Figure 17-17. The intersection point of the two lines represents the exact middle of the plywood you laid out.

Now measure up the minor axis from the center point half the width of the ellipse, which in this case is 4 feet. Set a temporary nail at this point; you'll use it to pull a radius.

Now comes the strange part, so follow closely. The radius that you pull from this point will be half of the major axis (or length) of the ellipse. If the ellipse will be 12 feet long, pull a 6-foot radius from this point down towards the major axis, as shown in Figure 17-17. You're trying to find the two places where this radius intersects with the major axis. You don't need to pull the whole radius as shown in Figure 17-17. All you really need is two marks where it intersects the major axis.

Once you've found these two marks, set a nail in each of them. At this point you should have three nails set in the plywood: one on the minor axis, and two on the major axis (equally spaced from the center intersection of the two axes). Tie one end of a string to one of the nails on the major axis. Loop it up and over the nail on the minor axis, and then tie it tight to the other nail on the major axis. Now you've established the length of string you need. The dotted line in Figure 17-18 represents the string pulled over the nails as described.

Remove the top nail that you set on the minor axis so it won't be in the way for the next step. Now mark out the ellipse, resting your pencil along the inside edge of the string, and slowly moving it along the string. Keep a constant pressure on the string with your pencil. You'll notice that the string will gracefully guide your pencil in a perfect ellipse. See Figure 17-18. When you've finished, you'll have scribed half an ellipse. You can repeat the process for the other side, or cut this one out and use it as a pattern.

After cutting the elliptical opening you've marked, nail it up in the opening in the ceiling drop. After it's up, you can establish where and how high

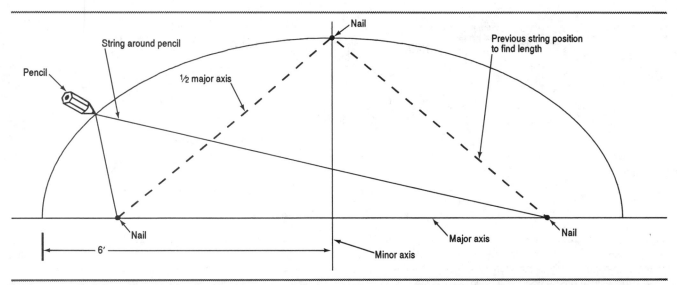

Figure 17-18 Once you've found the intersection points, place a nail in each one and another on the top point of your minor axis. Then tie a string from one major axis nail, over the minor axis nail, and tight to the other major axis nail. This is how you find the string length. The dotted line represents the string in this position. Now remove the nail on the minor axis. Hold your pencil inside the string and, keeping the string tight, swing the pencil around in an arc. The string's length will create a perfect ellipse.

you want the plywood ellipses that give the unit its rise into the ceiling. The height of the ellipse may be limited by framing members above it.

Notice in Figure 17-16 that three of the jack rafters had to be shaved a little on the underside to keep them from encroaching into the ellipse. But if jack rafters are all on the same plane, why would the underside of only three be in the way? Because the center ribs are cut to form an ellipse in width as well as lengthwise. In other words there are three ellipses here: an elliptical opening in the ceiling, an ellipse that rises into the ceiling three-dimensionally and follows the width of the opening, and finally an ellipse that rises into the ceiling three-dimensionally and follows the length of the opening.

As complicated as it sounds, it's really quite simple to lay out. First, find a section of floor as large as the elliptical opening in the ceiling. Snap out a major and a minor axis line that are perfectly square and perpendicular to each other. Pull 6 feet along the major axis line both ways from the center intersection point to copy the 12-foot length of ellipse that's in the ceiling already.

Now you need to establish the height of the highest point in the center of the ellipse, measured from the normal ceiling drop. If you run a straight-edge across the elliptical opening along the bottom of the normal ceiling drop, you can measure up to any framing member that appears to be in the way, and establish a height from there. You don't want to go too high in the center. For the ceiling to be effective, it should be only a slight rise. The ceiling shown in Figure 17-16 had about a 12-inch height in the center.

Once you know the height, you can return to those lines snapped on the floor. But you'll look at those lines from a different perspective now. The major axis line now represents the ceiling line and the minor axis line represents the height that the ellipse will rise three-dimensionally above the ceiling line.

If 12 inches of rise will work for your ceiling, measure up along the minor axis line 12 inches from the intersection point of the two lines. Set a nail at this point and continue to lay out an ellipse as shown before. Pull a radius that's half the length (6 feet) and find where it intersects with the major axis. Place two nails at these points, set up your string,

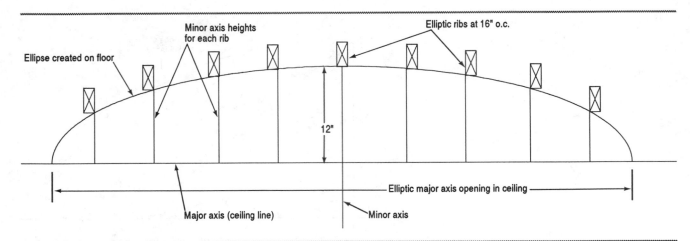

Figure 17-19   Here's a cutaway, sectional view of the ellipse as it's standing in the ceiling. The ribs are the individual members that make up the inner body of the drop. Once you know the height of the ellipse (in our case, 12 inches) you can use it as the minor axis to create this ellipse on the floor. The major axis is the same as the opening in the ceiling (in our case, 12 feet). When the rib members are laid out, take a measurement from the major axis to the high point they reach on the ellipse. This is the minor axis for creating this individual rib. The ellipse on the floor is just to find the minor axis for each inner rib.

and then pull the ellipse. The result is a cross-sectional view of the ellipse, as if someone cut it down the center.

With this view of the ellipse, you can lay out the rib members at 16 inches on center along the elliptical line you've drawn. Take a look at Figure 17-19.

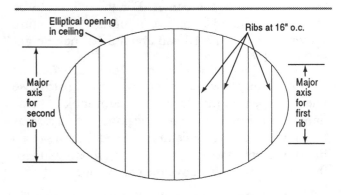

Figure 17-20   To find the major axis for the inner ribs, return to the opening in the ceiling. This is a view looking up into the elliptic opening in the drop. The lines represent the position of the ribs when they're installed. To find the major axis for the ribs, lay out the opening as shown. Then measure across the opening from one layout to the one opposite. This measurement is the major axis for this rib.

Notice the ribs drawn in at 16 inches on center that will frame in the body of the ellipse. You can lay out the ribs either of two ways: Start with one rib in the center and then lay out evenly both ways, or center a layout bay on the center mark and go from there.

Because you'll be creating a separate ellipse for each rib, it's best to keep them laid out evenly. Then you can use a rib on one side of center as a pattern for one on the other side. That eliminates a lot of time and labor, because you'll only have to lay out five ellipses instead of nine.

Once you have the layouts made, transfer them to the elliptical opening in the ceiling. You can find the major axis width for each individual elliptical rib and check that the rib won't interfere with any other rough framing in the ceiling. Figure 17-20 shows the opening laid out and how the ribs will appear when you're looking up at them. Notice how the widths change.

To lay out each rib member, you'll need the major and minor axis lengths. To find the major axis length, simply measure the length of the rib across the opening of the ellipse in the ceiling, from one layout mark to the other (Figure 17-20). Find the minor axis by taking a measurement from the ellipse you created on the floor. If you're dealing with the center rib, measure from the major axis line to the point where this layout falls on the ellipse drawn, which we know was 12 inches. Figure 17-19 shows where to take this measurement from.

Once you've cut a rib, check it for fit by simply holding it up in place along the layout marks on the elliptical opening in the ceiling. If it's cut properly, it will land directly on the cut edge of the ceiling ellipse when held on the layouts. If it's cut wrong, it won't fit the opening.

With the ribs all cut and checked for fit, nail them in place. Once the ribs are secure, install the blocks that will support the drywall or plaster wire. In Figure 17-16 the framer chose to use small blocks attached to each individual rib. Perhaps the jack rafters were too close to allow full blocks. He could have run a row of full-length blocks stretching from one rib to the next down the middle and possibly two more rows along the right side. This would have helped steady the whole structure and given the drywallers or plasterers a little more to nail to than the small blocks.

The Pay

The pay for drop ceilings is usually figured per room, or per drop if a room has more than one. A drop the size of the small closet we used as an example might pay around $20, since it should only take about an hour to build.

The sculpted drops pay a piece price per room. For the sculpt in the photographs, we figured the approximate time it would take, and then allowed $200 a day for a lead man, and $100 to $150 a day for his helper. After the first room we were able to

really fly. We provided all the saws, nail guns and compressors. The contractor supplied the nails. Of course, when we did this job the economy was healthy. There was lots of building going on with a resulting shortage of good carpenters. When building is slow, these prices may drop by a third or more.

Elliptical ceilings will pay by the hour, or pieced out per room. Figure out about how long you think it will take and set your price. Nobody holds the magic key to figuring out piece prices, especially for custom ceilings! If you're inexperienced and you tackle an ellipse ceiling that ends up taking a week, don't get mad at the general because you only made $5.00 an hour. On the other hand, if you did it in two days and will walk away with $100 an hour in your pocket, you may want to think again. How does the general feel about it? Maybe you should refigure your price and give a little bit back to him. This always helps for job security in the future.

The end of the job is never the time to renegotiate, as there will always be some hurt feelings on one side or the other. Talk to each other while the job is progressing. Just a simple "There's a lot more here than I expected. What do you think?" or "This is a piece of cake! How's it look?" will do. Keep your lines open. I've lost $6,000 on a contractor who wouldn't pay for extras, and I've walked away from $300 from a contractor who felt I did a piece job too quickly. Both instances could have been avoided through better communication. Both contractors approached me later with more work, but I refused to deal with either. Once a mistake, twice a fool, or something like that.

The point is both the contractor and the carpenter need to feel OK before, as well as after, the job. You as a carpenter are entitled to make good money if — and only if — you work fast and end up with a clean job. If you plod along and put up trash, you don't deserve the good money, *or* the next job.

Pick-up Work

Inevitably, during the building of a house, there are mistakes, changes and assorted mismatches that are best put off until later. No matter how high the quality of work that goes into a building, there's no escape from having at least a few things to correct once the house is substantially complete. Of course, a well-organized and experienced crew will leave less work at the end of the process. But whether it's through greed or ignorance, tract crews seem to breed this mop-up work — usually referred to as *pick-up*. I think it got its name from that famous phrase "You can *pick up* your last check when this house passes inspection."

It's hard to define exactly what pick-up is. What might be considered piece work on a tract would be considered pick-up for an hourly crew. Especially on large customs, pick-up work can be very elaborate. The skylight/ceiling drop framing shown in Figure 18-1 is a good example. Where one contractor might piece this out, another will use it to keep his own guys busy.

Depending on the organization (or disorganization) of the job, pick-up may consist of a few missing fire blocks, a bathtub platform and some stud and door straightening. Or it may be three weeks of wall straightening, drop remodeling, skylight readjusting, shear panel renailing, fireplace leveling, or row after row of fire-block fixing.

On a tract you'll usually have a single crew doing pick-up at an hourly rate. On customs the company may have one or two really good pick-up carpenters that do all their houses, or the regular carpenters may do it. On tracts and customs alike I've done pick-up on a piece price, but only when a company that knew and trusted me was in a bind to get the work done by a deadline.

Figure 18-1   This framing could actually be called a drop, but it's typical of some of the more cut-up pick-up work you might encounter on a custom house.

months would teach you more than a year at a trade school — and you'd be paid for it! Not everyone is able to jump right into piece framing. Pick-up carpentry has been a good stepping stone for countless beginners. There's no better way to figure out how something works than to have to fix it when it's broken.

Two Types of Pick-up

There are basically two types of pick-up work: repairing or changing something done wrong, or completing a job that wasn't accessible until now.

The first step toward fixing a problem is realizing that it exists. That's the responsibility of the supervisor. If he paid someone to plumb and line a building and finds out later that it wasn't done properly — and the person he paid is long gone — he'll have to pay to have it fixed. Before a pick-up crew starts a house, the supervisor will go through the house and mark out each item that needs fixing or finishing. Frequently, on a tract, each model will have the exact same problems. Once you've done a couple of houses, there's no need for the supervisor to walk the remaining houses until you're done.

Pick-up carpentry can be a depressing job. Sometimes it seems that instead of fixing some problem it would be better to tear down the whole house and begin again. And that's probably been done before! To do pick-up work on a tract, it helps to have an optimistic, happy-go-lucky type of personality.

Pick-up carpentry is where a lot of carpenters begin framing. Working alongside an experienced carpenter doing pick-up at $10 an hour for six

Problems that weren't apparent in the beginning have a way of spreading throughout a tract, from house to house, like some strange virus. They're also cumulative in nature. For example, if the plumb and line carpenter braced up a kitchen wall ¾ inch out of plumb, it only takes a few minutes to undo the brace and then replumb it.

But let's say the joisters got in there before the problem was identified and ran their joists along the top of the wall, securing it ¾ inch out of plumb. And then maybe it was a shear wall with A35s every 16 inches from the top plate into the floor blocks.

Then, the day after inspection, the cabinet installer throws his level on all the kitchen walls. Without plumb walls his cabinets are going to be a real problem to install. He discovers the out-of-plumb wall and refuses to hang any cabinets until it's fixed. So now a job that originally should have taken minutes to fix will take the pick-up carpenters a couple hours at the very least. On a bad day they might damage some plumbing or electrical work in the process, which will take a few more hours for the other subs to come fix.

As you can see, these small errors have a snowball effect. A capable foreman can keep these situations from getting this far out of hand. It's up to him to find workers he can trust not to leave him holding the bag at the end of the job. Problems are always going to occur; it's unavoidable. But the more problems *you* leave behind, the more your reputation (and wallet) are going to suffer.

■ Fixing Door and Window Openings

A typical problem area that needs attention during pick-up is rough openings for windows and doors. The super will check all the windows against the window schedule for correct rough openings. If they're not right, they need to be fixed. If the opening is too big, no problem. You just add on what's needed to make the opening the right size.

If it's too small, you might have big problems. If the wall is a bearing wall, you can't remove the trimmers, so you'll probably need to install a larger header. If it's not a bearing wall, you might be able to remove a trimmer and replace it with thinner material to gain the room you need.

Doors are checked for proper rough openings and also for plumb. Typically, an interior door should have a rough opening that's 2 inches wider than the door callout. If the door is a 2′6″, the rough

opening would be 32 inches. Exterior doors have 1¼-inch-thick jamb stock, so they need a rough opening that's around 3 inches wider than the callout.

When you check a door opening for plumb, you'll want to check each trimmer both ways. That is, check the trimmer for level along its edge, as well as on its face. If it's out of plumb a lot (let's say 1 inch), try adjusting both trimmers to plumb and then check your rough opening. It might be too small. If it is, you'll have to move the bottom of one of the king studs over and cut off the excess bottom plate with a Sawzall.

Now check the edge of the trimmer. If it's out, look down to the bottom plate and see if it's on the snapped line. If it isn't, try to persuade the bottom plate over with your hammer, then recheck it. If both trimmers aren't plumb and on the line when you finish, don't expect the finish carpenters to be sporting you any coffee at break time.

■ Hanging Doors and Windows

A rough framing bid almost always includes setting windows and exterior doors. This is because the finish carpenters don't usually come around until after the walls and the roof of the house have been sealed. Of course, this wouldn't count any custom fixed glass unless there were previous arrangements. We're only talking about delivered windows that you set into prebuilt rough openings.

The first step in setting windows is to unwrap them from their miles of cardboard and plastic, then spread them around the job site at the right openings. You'll need a level, a hand staple gun, a finish nail gun, some cedar shims, and some 6-inch sisal kraft paper. For wood windows with plaster mold attached to them, you'll want to wrap the whole opening with the paper. For wood or aluminum windows with nailing fins, wrap the bottom and two sides but leave the paper off the top until the window goes in. Then put it over the top window nail fin.

When you wrap the opening with paper, start along the bottom sill. Let the paper go wild past both trimmers 6 inches or so. Then lap the side pieces over these excess bottom pieces. Let the side pieces run long above the header so the top can lap and cover it. Then the top piece goes over the sides. Why do you work your way up from the bottom? So if any water makes its way in, it will drain down the paper. The way the paper is lapped should keep the water from getting under the paper and in around the window.

When hanging the paper for windows with nailing fins, you can run the paper evenly along the edge of the trimmer without encroaching into the rough opening. For windows with plaster mold, let the paper hang in an inch or so. Then when you set the window you force it in past the paper, causing the paper to bend in. That gives the window a little extra protection.

Once the paper is stapled around an opening, you can set the window in place. Then put a level on the sill and adjust it with cedar shims. When the bottom is level, nail it off (the bottom only) either with hand nails through the nailing fin, or with a finish nailer through the plaster mold. Always use galvanized nails!

Now you can adjust it for plumb and square. One easy way to do it is to sight the edge of the movable sash with the middle bar on the fixed sash. Open the window slightly until the two middle bars on the sashes are aligned. If the window is out of square, you'll see a noticeable difference between the sashes. You can adjust this with cedar shims along one top edge. When the gap is even, you know that the window is square and you can nail it off completely.

Hanging a pre-hung exterior door is very similar to hanging windows. Get the door in place and level up the hinge side with cedar shims. Once it's good, nail it off with 16d or 8d galvanized finish nails.

Now adjust the jamb on the knob side with shims until it meets with the door edge evenly from top to bottom. Use the door edge as a straightedge to sight the jamb with. There should be a gap about 1/8 inch between the door and the jamb. When it's even, nail it off with at least six nails: two on the bottom, two in the middle, and two on the top.

■ Dealing with Loose Ends

You can continue with the second type of pick-up work, which is installing and checking items that weren't available to work on until now. This phase of pick-up carpentry is a matter of buttoning up the loose ends in the house.

When the supervisor checks out a building, he'll either make a *punch list* on a piece of paper, or write a note on the floor with a keel with an arrow pointing to the problem or the addition. I like using the keel for two reasons. First, a piece of paper can get lost. Second, carpenters can spend all day walking back and forth to wherever the list is posted. If everything is listed right on the floor, they can stay in one room at a time and work on each item until the room is done.

Let's take a bathroom for an example. The biggest pick-up project might be a tub platform. The supervisor will usually snap out the lines for the supporting walls and steps and note the proper stud sizes and the plywood thickness. That's all you need for a simple platform.

Next you might notice a note for a clip wall to separate the toilet from the tub. A clip wall is a short wall, usually only 30 to 48 inches high. Again, the supervisor would snap out the lines or write out a dimension and stud size for you on a neighboring wall, and mark where it should end.

You might find a note that reads "Backing missing, look up." Looking up you'd notice a piece of ceiling backing that was left out. Or you may find a note that reads "Need hangers on head-out." Pick-up carpentry is mostly an array of simple little jobs like this.

Fireplace hearths and structures built around zero clearance fireplaces are also typical pick-up jobs. There are usually so many braces in the way that there's no chance to get to it until the end. The term *zero clearance* is a bit misleading. Typically you have clearances from 1/2 inch to 2 inches along the top, back, and sides.

Figure 18-2 shows a typical zero clearance fireplace and the framing installed around it. The walls above the fireplace were hung off of the scissor trusses. Notice the pipe rising up behind the fireplace. This is a flue from a lower fireplace. So this structure surrounds the upstairs fireplace and acts as a flue chase for the one downstairs.

Any flue chase or open column that's longer than 10 feet usually has to have a fire stop placed every 10 feet or so, depending on your local code. Most tall columns are built with supporting blocks every 8 or 10 feet. You can use these for fire blocks to accompany the fire stop. The blocks keep fire from going up between the studs, and the fire stop keeps the fire from spreading up the open area inside the column.

The blocks need to all be lined up in order to make a good fire seal. If they weren't put in evenly, you'll either need to fix them or install new ones before you add the plywood. When adding the plywood, install a rim of 2 x 4 blocks around the inside of the column to support the plywood. These fire stops can take a lot of time, especially if the electrician has already loaded up the column with wires.

For the flue chase in Figure 18-2, we headed out the floor joists to allow for the flue to pass through, but we sheathed over the head-out when the floor was sheathed. Later, when the flue was installed, we cut a hole in the sheathing for the exact size needed. Had we left the head-out open, we would have had to fill in the open spaces around the pipe to create a fire stop.

In some parts of the country you're required to install fire stops in hall drops that are over 10 feet long. They keep fire from racing down the chutes formed by this framing. The fire stop is made by sealing the drop solidly with scrap plywood every 10 feet.

Another typical pick-up job is the platform for the hot water heater. Generally a water heater must be kept 18 inches off the floor. If you have a whole closet devoted to the water heater and furnace, you can build a platform that fills the closet. You'll need fire blocks set halfway exposed above the platform floor level. With this kind of setup, there's usually a set of double doors installed. Just build the platform on the inside of the closet, and leave the

Figure 18-2   Framing around zero clearance fireplaces is a typical pick-up job. This framing in particular was hung from the lower chord on the scissor trusses.

trimmers open to accept the doors. This gives a clean look and is more economical because you can use common full-length doors. Sometimes you'll just build a little 18-inch-tall platform in the corner of a garage for the water heater.

Straightening Studs

The last step in pick-up is usually wall straightening. This is a task that takes patience and wit. There are two popular methods to straighten a stud. I call them the *band-aid method* and the *strongback method*.

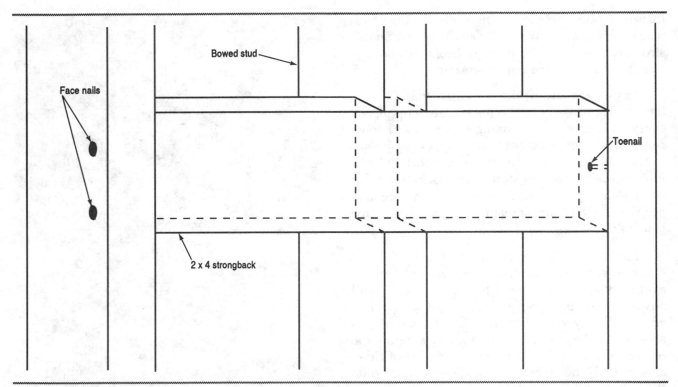

Figure 18-3   A strongback is a 2 x 4 that's set flat into a bowed stud, spanning to the studs on both sides of the bowed stud. With one side nailed to a stud, the other side can be controlled with a toenail until the stud is straight. Then the second side can be face nailed securely.

Most of you are familiar with the band-aid method. It's illustrated in many books on construction and goes something like this: Find the bent stud, put a cut half way through the back side, stick a shim in the cut until the stud lines up with its neighbors, and then nail a plywood gusset (or band-aid) on both side faces of the stud to hide the cut and to keep the stud in place. This method generally works with 2 x 4 studs, but it reduces the integrity of the wood so much that most inspectors frown on it. For 2 x 6 to 2 x 10 studs, forget it! You end up cutting three quarters of the way through the stud before you get any movement.

For straightening studs, I like to use the strongback method. See Figure 18-3. It's easy and very controllable. I usually work in one room at a time when straightening studs. I'll go through the room with an 8-foot level or straightedge and lay it horizontally in the middle of the wall across six studs or so. If there's a bowed stud, the level will rock back and forth on it. Once you've determined which stud is bent, draw a line on its face with your keel and then move down the wall. Check the whole room in this manner and mark out all the bowed studs.

When the studs are marked, you can notch them for the strongback. A strongback is a 2 x 4 set flat into the wall that's centered on the bowed stud and stretches to the studs on both sides of the bowed one. Start by making a cut, 3½ inches wide by 1½ inches deep, into the face of the stud, about midway up its length. The tighter you make the cuts, the stronger the stud's going to be, and the happier the inspector will be when he sees it.

Once the notch is made, take a measurement between the studs on both sides of the bowed one. If all the studs are on 16-inch layout, the strongback will be 30½ inches long. Cut the strongback and place it in the slot between the studs and face nail it to the bowed stud. Then nail it to one of the other studs. Face nail through the stud into the block, making sure to keep the face of the block even with the face of the stud. Now push the strongback in to meet the face of the other stud. You'll straighten out the bowed stud.

Try using a toenail through the face of the strongback and into the stud. Drive the nail in a ways and then check the stud with the level again. If it's still rocking, send the toenail in a little more and then check with the level again. If you go too far, you can hit the back side of the strongback and bring it back out a little. Using a toenail like this makes for an easily-adjustable setup. Once you have the stud straight, face nail through the stud that you were toenailing into and on into the strongback. The face nails will keep the toenail from slipping back out later.

The Pay

Pick-up work is steady and that's the type of work a lot of people prefer. Since you're doing something different all the time, it's hard to figure piece prices. But that's not to say it can't be done. For a while, three of us were doing all the pick-up and the drop ceilings at a rate of $600 per house. It all depends on whether you and the supervisor can work out something that's comfortable for both. But most pick-up is done at an hourly rate. It pays anywhere from $5.00 to $18.00 an hour, depending on your experience.

So there you have it. We've built a whole house within this book. Well, maybe we left out a nail or two, but for the life of me I couldn't tell you where. And that's about how carpentry is. You do your best to remember what you've been taught, but you'll never remember it all. All the lessons and teachers you've had will blend together until you come up with your own way of doing things.

I've only included *my* way of doing things. Although I'm partial to it, I won't tell you it's the only way. It's what works for me. It's what has kept me in money through all the economic ups and downs. And if you stay in this business for any length of time, you'll find out that's what takes the most skill. You may be able to do cartwheels on a 20-foot top plate while drinking coffee and dragging a stack of trusses. But if you can't scare up work in slow times, it doesn't mean a thing.

Don't be afraid to adjust your prices when things slow down. You'll be forced to if you want to keep busy! But if you understand most of what we've done in this book, don't underestimate your skills. There are countless remodel jobs out there when times get slow. You have the skills and the tools. There's good money to be made in little jobs. Get your name out there. Treat people right and they'll do the same. But most importantly, always do clean work — even if it costs you more in the short run. In the long run, all you really have is your name, so keep it clean. And try to have some fun in the process.

Glossary

A

"A" brace - A temporary brace that holds a wall in place until another wall can be built and attached to the first wall, holding them both in place. The brace is shaped like the letter A, with two side arms and a middle support that keeps the sides from spreading. See Figure 5-10.

A35 - A hardware clip with a 45-degree angle, commonly attached to the top plate and rafter block to transfer shear from a wall to a roof. See Figure 3-5.

B

Balloon stud - A stud in a balloon wall. Balloon framing uses one-piece studs that extend from the foundation to the roof to form the walls of both stories in a two-story structure. A balloon stud is taller than the typical stud height. See Figure 5-12.

Balloon wall - Any wall that is taller than the typical wall height throughout the building and has a flat top (as opposed to the sloped top of a rake wall). Walls in areas around stairways, for example, are balloon walls at least two stories tall. See Figure 5-12.

Barge rafter - The outside rafter on a gable roof end, known as the barge board, barge rafter or fly rafter. It is fastened to the roof sheathing above and the lookout blocks which sit behind the rafter. Typically, it overhangs the outside wall a few feet. The barge rafter is also the gable fascia board. See Figure 14-8.

Bearing wall - A wall that carries a floor or roof load as well as providing a partition between rooms. Bearing walls are essential for the stability of the structure. Interior bearing walls are designed with an enlarged concrete footing to handle the extra weight they will be carrying. They also have larger door and window headers than non-bearing walls because an alternate load path must be established to carry the loads across the top of the openings.

Birdsmouth - The triangular cutout on a rafter that allows it to sit on the top plate of the outside wall. A seat cut and a plumb cut combine to make up a birdsmouth. See Figures 12-1, 12-3 and 12-28.

Bottom plate - The lowest horizontal wood member in a wall. The bottom plate rests on the floor and is attached to the slab with bolts, or hardened pins. The bottoms of the wall framing studs are nailed to it. This is also called a sole plate.

Buck up - Nailing the upper cripples to the headers and the lower cripples to the subsills around door and window openings before you begin framing the walls. Each door and window is prepackaged and ready to integrate in the wall. Before building the walls, the person doing the cutting cuts the headers, subsills and top and bottom cripples and places them adjacent to their locations in the finished wall. These are nailed in first (bucked up), then the wall framing is completed. See Figure 5-1.

C

California fill - The art of stacking one roof onto another, and creating two valleys, as opposed to stacking the same section using two (structural) valleys. The roof to be stacked upon is usually sheathed first, to transfer the shear and to create a surface to work on. See Figures 11-12 and 13-29.

Cat's paw - A pry bar with notched and flattened ends for pulling nails. One end is bent at a right angle to the bar.

Catwalk - A board that runs horizontally on top of ceiling joists or the bottom chord of trusses, extending from one outside wall of the building to the other. It ties the joists together and holds them on layout. See Figures 7-23 and 11-22.

Ceiling joist (CJ) - The horizontal structural members spanning the top wall plates and to which the ceiling is attached. Ceiling joists tie the tops of the walls together so that the downward weight of the roof does not push the tops of the walls outward. They redirect the load so that, instead of pushing out, the load goes down through the walls to the foundation. Ceiling joists are sized according to the length of the unsupported distance they must span and the load they must carry. See Figure 7-22.

Channel - The framer creates a channel where two walls intersect. The channel provides a stud to nail the intersecting wall to and backing for the drywall at the inside corners on either side of the intersection. There are two types of channels; an outside corner channel (where two corners meet) and an interior channel (where a wall intersects the middle of another wall). See Figures 4-14, 4-15 and 4-21.

Cheek cut - A beveled cut on the end of any rafter that allows it to be placed flush against a diagonally intersecting member, usually a valley or a hip rafter. See Figure 12-3.

Concrete slab - A flat, horizontal concrete pour, often the floor of a building. See Figure 4-1.

Conventional roof - A roof that is calculated and cut entirely on the job site. No trusses are used. See Figures 12-2, 12-4 and 12-6.

Cripples - Short wood members that run vertically above and below a window opening or above a door from the header to the top plate.

Crown - The outward curve of a board visible if you sight it edgewise. Some boards have severe curves while others are very slight. As a rule, boards should be installed with the crowned edge up. When they're loaded, they will deflect slightly and become straight.

Custom home - A single home that's custom designed and built for clients on their lot. It is usually one of a kind. See Figure 1-1.

D

Detail - The instructions written out on the top and bottom plates for the wall builders. Door and window headers, cripples, studs, posts, channels, and fire blocks are all detailed in by the "detail man." See Figures 4-12 and 4-23.

Developer - A person, group of people, or a corporation that plans and finances construction projects for profit. Developers' projects come in all sizes, from large housing tracts and commercial/retail complexes to a single family home.

Doubler - Any two structural members that are nailed together as one. Floor joists are commonly doubled up to give support to second floor walls that will bear directly on the doubler. First floors of multilevel apartment buildings are sometimes double 2 x 12s at 16 inches on center. See Figure 7-8.

Drop ceiling - A false ceiling that is built below the already-installed ceiling. This is done for aesthetic reasons, such as for the installation of lighting panels, or to create a concealed space to house air ducts, plumbing, or electrical connections. See Figures 17-2 and 17-3.

E

Ear - When a double top plate is installed, the interior channels are left open to accept the ears, or laps from other walls. Walls that don't have corner channels will have an extra 3½-inch ear to lap the corner wall it meets. See Figure 5-8.

Elevations - Exterior decorations that are mounted on the walls during framing. They may include built-up bands around windows, arches, pot shelves, bay windows, and other frills. See Figures 16-1 and 16-2.

Exposed - Framing members that will be visible after the structure is completed. Exposed material is usually painted rather than being covered with stucco as are other rough framing members. See Figures 7-3 and 15-7.

F

False tails - Fancy beams that are mounted to every other rafter around the perimeter of a roof to give the illusion that the rafters are larger than they are. They are typically corbeled or rounded on their ends. See Figures 14-18 and 14-19.

Fascia - A board, attached to the ends of the rafter tails, that creates the exterior finished edge around a roof. See Figures 12-4, 14-2 and 15-2.

Finish carpentry - The finish work on a building which includes hanging the doors, installing the trim and molding, cabinetry, and any other woodwork on the interior of the building. Finish carpentry is done after the drywall is hung, and must be done to finer tolerances than rough carpentry, as the work will be exposed in the finished structure.

Fire block - A block, usually placed horizontally or at an angle between studs, which is built into a wall to slow down the spread of fire. The block must be same size as the studs so there are no gaps once the drywall is installed. They block the spread of fire by limiting the open airspace within the framing. Typical locations for fire blocks are at the rim of a drop ceiling, along a stairway, and adjacent to any draft stop up a fireplace shaft. See Figures 5-17 and 5-20.

Float backing - Backing attached to the bottom of floor joists, ceiling joists or trusses, which moves with the joists as a load is placed on them. If an interior wall dips down due to a dip in the concrete, the ceiling backing will need to float or the ceiling will show the same dip. See Figure 11-18.

Flying hip/valley - A hip/valley that runs from one ridge to another. A normal hip/valley extends from the top plate to a ridge. See Figure 13-1.

Framing foreman - The person in charge of all the framers on the job.

Frieze block - Blocks mounted between rafters where they attach to the walls. They are installed at a square 90 degrees with the top of the rafter, with the bottom edge of the block against the wall. The stucco or siding will then come up the wall and end at the bottom of the frieze block. See Figure 13-32.

G

Gable - A roof with a uniform pitch from the ridge to the outside wall. All the rafters are cut to the same length and joined in the center to form a peak, with the two sides of the roof sloping down from that peak. The gable ends are the triangular ends of a building with a gable roof, between the top wall plate and the ridge of the roof. See Figure 12-5.

Galvanizing - The process of applying a protective zinc coating to iron. The zinc inhibits corrosion and oxidation of the iron. Any exterior exposed wood, whether it is stained, painted, or left natural, should be installed using galvanized nails. Nongalvanized nails will rust and bleed, causing long, dark, unsightly streaks down the material.

Girder - A structural beam used to support concentrated loads at points along its length. Girders are typically used to reduce the span of floor joists in buildings with a first floor built of wood rather than a concrete slab.

H

Hang nail - A nail that is set halfway into the top of a board and then bent over to create a hanger that allows a board to be hung there temporarily. For example, backing can be hung from the top plate, then pulled up when you're on the joists.

Header - The beam that runs horizontally above a door or window opening, supporting the floor, ceiling, or roof that's bearing on the top plate above it. The header redirects the load and adds support between the studs. See Figure 5-4.

Head-out - A cutout in a section of joist to make room for another material. Two small joists, running perpendicular to the cut joist, are then added to support each cut end. Floor head-outs are installed if a joist is in the way of plumbing. Rafters are also headed-out to allow room for fireplace flues. Lastly, floor joists are also headed-out to attach a horse for a flight of stairs. See Figures 7-10b, 9-7a, 9-7b and 9-7c.

Heat bay - A span left between joists, top cripples, or studs to allow a heat duct or vent to be mounted in a specific location. Top cripples may have to be centered over a door to allow room for a heating vent to be placed there. This would be called a 14½-inch heat bay above the door.

Hip - A roof style that has sloping sides and ends so that the eave line formed is constant on all walls. A bastard hip is a single hip in a roof that is at a different pitch than the other hips. See Figure 12-2.

Hip rafter - The main structural support on a hip roof. See Figure 12-2.

Hold-down (H.D.) - Hardware used to attach structural shear walls to the slab in earthquake-prone environments. Hold-downs are bolted to posts that are built into the wall, and attached to bolts that are set into the concrete slab. See Figures 3-5 and 10-2.

Horses - The diagonal supports for a set of stairs. Also called stair stringers. See Figures 9-7 and 9-13.

J

Jack rafter - A rafter that spans from the wall top plate to a hip or from a valley to a ridge. See Figure 12-2.

Joists - Horizontal structural members that support the floor or ceiling. See Figures 7-8, 7-13 and 7-14.

Journeyman - A tradesman who has the experience required to complete any task in his chosen field without supervision. A journeyman is also qualified to oversee a crew of people if needed. At one time, a union apprentice had to serve a certain number of years to qualify as a journeyman. Today, the term is often used to designate competence.

K

Keel - A large crayon used to write on lumber and concrete. Frequently a carpenter's pencil will barely show up on wood or concrete, but a keel mark will appear bright and last through many rainstorms.

King rafter - A rafter that's the same size and cut as a common rafter, but which may not be found on layout with the rest of the body of rafters. The exact center of a king rafter must be aligned with the center point of the ridge-hip connection. This ridge-hip connection, with two cheek cuts on each hip, can't be adjusted, so many framers avoid using king rafters. See Figure 12-30b.

King studs - The studs on either side of a window or door to which the header is nailed. See Figure 5-4.

L

Landing - A level platform between two flights of stairs designed to break the climb or allow for a change in direction. The floor areas at the top and bottom of the stairs are also called landings. Typically, a landing is as wide as the stair treads, or a minimum depth of 36 inches (after drywall). See Figure 9-19.

Layout - The numerical pattern or standard measurement set for the installation of individual studs, rafters or joists. Layout is commonly called out as "on center," meaning each member is placed at a constant distance from center to center. A 16- or 24-inch on-center layout is typical for a rough framing layout. With these standard layouts, plywood, drywall, and interior wall paneling will easily mount to the studs without any cutting. Layout is the golden rule to organized and efficient framing.

Let-in brace - A 1 x 6 brace that is set into (or let-in) notched studs so that it fits flush to the studs. By making angled ¾-inch-deep notches in the studs, the brace is buried below the plane of the wall. When the wall is built the brace is tacked in place with 8d nails. Once the wall has been plumbed, the nails are sunk, ensuring that the wall is solid. See Figure 5-7a.

Level - Aligned and true on a single horizontal plane. Also the name for the tool used to check this. See Figure 6-1.

Lower cripples - Short vertical wood members that run from the subsill to the bottom plate below a window opening. They're spaced 16 inches on center.

M

Miter cut - An angled cut used to join two boards together. Each board receives a mitered cut of the same angle along its face. Then they're lapped to create an almost invisible joint. This type of cut is used primarily for exposed wood, such as fascia or molding.

N

Non-bearing wall - An interior or exterior wall that does not support any additional weight (such as a floor or roof), but serves only to divide or partition-off a room. A non-bearing wall is supported beneath by just the thickness of the slab, which is 3 to 6 inches thick. Bearing walls have an enlarged concrete footing to help them support the extra weight that they carry.

Nut and shoot - The first step in plumbing and lining walls, which consists of attaching and tightening a washer and nut to all the bolts sticking up through the bottom plate, and then shooting the remaining walls to the floor with a powder-actuated nailer. Before the bottom plate is tightened or shot in place, it must be moved so that it is aligned exactly between the snapped lines. See Figures 6-2 and 6-3.

O

On center (O.C.) - The measurement from the center of one structural element to the center of the next. Blueprints commonly use the abbreviation O.C. when detailing a layout.

Outlookers - A 2 x 4 that's laid flat and let into the gable truss or rafter so that its top is flush with the top of the gable. It is cantilevered from the second rafter out through the gable to support the barge rafter. Since outlookers are exposed, the material used for them is usually selected for appearance. See Figure 14-8.

Overspan - A span larger than the plans call for. An overspan is created when a member, such as a joist or rafter, is installed with a bay that's larger than normal. For a 16-inch O.C. layout, a normal bay is 14½ inches wide. If for some reason it is necessary to install a joist with a 17-inch bay (perhaps in order to avoid a plumber's drain pipe), it would create an overspan. A few overspans aren't cause for alarm, but don't overuse them. Inspectors don't like to see lots of overspans. If the overspans are all to code you can't be cited on their account. However, if the inspector disapproves of their number he's likely to look at the rest of the work with a more critical eye. See Figure 7-1.

P

Per square foot - Piece pricing in which a task is given a set price per square foot of building or material installation. This is a common way to pay for carpentry work on tract housing. A piece price remains the same for as many houses as are available to be worked on. For instance, a 2,000 square foot house may pay 15 cents a foot to snap, plate and detail. This works out to $300 a house. To sheath the roof may pay 10 cents per square foot of sheathing. If the house takes 75 4 x 8 sheets, it would pay $240. A whole house will cost from $3 to $10 a square foot to be rough framed. Garages are rarely figured in the overall square footage price.

Pettibone - A large rough-terrain forklift commonly used to move material around on a job. See Figures 5-13 and 7-4.

Pick-up - Work that is left over after the majority of the house is framed. This can range from mistakes that need to be corrected to tasks that could not be completed until construction has reached this point.

Piece price - A common method used to pay carpenters on a tract and on some custom work. The carpenter is paid a set price to complete a given amount of work, regardless of how long it may take. Carpenters can make as much as $1,500 a week piecing, or as little as $30 for a week's work. (I have honestly witnessed a man make $30 for a week's work piecing!)

Plate - The bottom and top horizontal structural members to which the wall studs are nailed. There is one bottom plate and two top plates. See Figure 4-17.

Plating - The second step in snapping, plating and detailing the wall framing. Once the floor plan has been snapped out on the slab, the top and bottom plates for each wall are cut and laid in place, with the top stacked on the bottom. This is known as plating. Walls with bolts need the bottom plate drilled and then laid in place. Once all the wall groups that make up the house have been cut, drilled and put in place (plated), the detail person transfers the door, window, post and other measurements from the plans directly onto the plates. The wall framers use these detail marks as guideposts for their work. See Figures 4-20 and 4-23.

Plumb - True and level on a vertical plane. A level placed along the outside stud in a wall can be used to check the wall for plumb. See Figure 6-1.

Plumb and line - When the entire group of walls in a house have been framed, they need to be plumbed up, temporarily braced, and sighted to make sure the long walls are straight. This temporary bracing up and checking for a straight wall line is referred to as plumb and line. See Figures 6-5, 6-7, 6-9 and 6-11.

Plumb bob - A weight, usually with a point at the bottom, that is suspended from a string and allowed to swing free. The force of gravity causes the string to hang in the true vertical plane. By holding the string against the top plate and letting the bob hang free just above the top of the bottom plate, the plumb bob can be used to align the framing for high walls. See Figure 6-13.

Pneumatic - Powered by compressed air.

Pressure block - A block nailed against a rim in every other bay. The joists are face nailed to it and toenailed to the rim. A pressure block can be used in place of a hanger in some circumstances. See Figure 17-4.

R

Rafter - The structural support members for the roof of a building, extending from the peak of the roof to the top plate of a wall. See Figure 12-2.

Rake wall - A wall that is sloped along the top, with a high point and a short point. Often a shear wall will continue from the floor all the way to the roof sheathing, so the top of the wall is angled to match the angle of the roof. See Figure 5-21.

Ridge - The highest point of a sloped roof. The ridge board is the upper support member against which the tops of two opposing rafters rest. See Figures 12-2, 12-3 and 12-4.

Rim - The outermost joist on a floor. The rims are situated directly over the exterior walls. Rims also surround a stairway opening, or any other opening in a floor. See Figure 7-17.

Rise - The vertical distance that a flight of stairs rises or a single step in a staircase rises. Also the vertical distance from the top plate on a wall to the ridge of the roof. See Figure 9-5.

Rough carpentry - One of the two main carpentry specialties. Rough carpenters complete the structural framework and the major parts of the building. Finish carpenters, the other main carpentry field, do the remaining work, such as hanging doors and installing molding and trimwork, after the interior wall finish (drywall or plastering) is completed.

Rough opening (R.O.) - The rough-framed opening in a building wall intended for the installation of a window or door. The dimensions are provided by the window or door manufacturer. For a window, the R.O. is usually ½ inch larger than the actual window in both directions, to leave room for adjustments. The R.O. for an interior door is 2 inches wider than the door: 1½ inches for the ¾-inch jamb on either side and another ½ inch for adjustments. Exterior doors need 3 inches, as they usually use 1¼-inch jamb stock to support the solid core doors.

RS1S2E - An abbreviation for "Resawn one side and two edges" which is typical for fascia and other exposed exterior wood.

Run - The horizontal distance covered by an entire flight of stairs. Also an individual tread, the horizontal portion of one step, in a flight of stairs. Lastly, the horizontal distance covered by the roof from the top plate of the wall to the midpoint, or one-half of a roof's span. See Figures 9-3 and 9-5.

S

Scissor truss - A truss with a steep exterior slope and a bottom chord that angles up in the middle. This creates, on the interior, a sloped ceiling surface for either a vaulted or a cathedral ceiling. See Figure 11-1.

Seat cut - The horizontal cut that, in combination with a plumb cut, forms a birdsmouth cut in a rafter. The birdsmouth allows a tight, flush fit to be made between the rafter and the top plate. See Figures 12-1 and 12-28.

Shadow board - A board used to hold up the first row of roof tile. Without it the first row will dip an inch or so since it isn't lapping like all the tiles above it. At one time a shadow board was mounted along the top edge of a fascia to create a shadow along the fascia for aesthetic purposes, which explains how it came to have its present name. See Figure 14-2.

Shear panel - A plywood panel that is mounted on a wall from the bottom plate to the top plate with all the edges blocked. It is installed to provide shear strength and is nailed according to the shear schedule provided in the plans. See Figure 10-4.

Snapping - Laying out the floor plan on the concrete slab using a chalk line. This is the very first step to framing a house. Each wall is snapped in place with chalk exactly as the blueprints show it on the floor plan. See Figure 4-3.

Spec house - Short for speculative house. A spec house is built as an investment to sell rather than for a client to live in. The person fronting the money is "speculating" that there is a profit to be made if he can build the house and sell it quickly.

Speed square - Originally manufactured by the Swanson Company, the speed square is cast in a triangle with one edge fattened with a lip. By placing the lip along the edge of a board, you can swing the square around and hold it at any angle you want. It reads in degrees, common roof pitches, and hip/valley pitches. See Figure 12-15.

Stacking - The act of putting trusses or rafters in place on a roof. A crew that is building (not cutting) a roof is said to be stacking the roof, and each crew member is called a stacker. See Figures 11-2 and 13-3.

Stair gauges - Moveable indicators that fasten onto a framing square. They're used to mark two measurements that are used again and again. These gauges are most commonly used in stair cutting, to mark out rise and run on the stringers. See Figure 9-9.

Stand - The height that the rafter sits above the wall, measured from the seat cut, up the wall line to the top of the rafter. It is also known as the throat. See Figures 12-1 and 12-18.

Starter board - Resawn 1 x 8s, commonly referred to as starter board, used to sheath the rafter tails in an exposed overhang. See Figures 14-2 and 15-2.

Stick framing - Building a wall in place, one stud (or stick) at a time. The bottom plate is bolted down, then the two outside studs are installed to support the top plate. All the other studs are raised in place. In some parts of the country the whole house is stick framed.

Story pole - A pole, with measurements marked on it, used to double-check a stair calculation or some other standard measurement. See Figure 9-5.

Stringer - The main diagonal support member on which stair treads rest, also called horses. See Figures 9-7 and 9-13.

Strongback - A 2 x 4 let flat into the face of a bowed stud to straighten it. See Figure 18-3.

Stud - Vertical structural member, usually a 2 x 4 or 2 x 6, that makes up most of a framed wall. Studs are typically spaced 16 inches on center throughout a wall in between windows and doors. The term stud usually refers to the boards that are delivered precut. For 8-foot walls, the studs are 92¼ inches. Combined with three plates, the walls are 97 inches tall when finished. See Figure 5-3.

Subsill - The bottom, horizontal member in a window opening that runs parallel to the header and supports the actual sill of the window. See Figure 5-4.

Supervisor - The person, under the general contractor, who is in charge of all the subcontractors on the job.

T

Template - A pattern cut on a board that can be traced onto other boards. For example, the detail for a rafter tail and birdsmouth could be made on a smaller piece of wood the same width as the rafter material, then used as a guide to trace the cut on each rafter. See Figure 12-25.

Throat - The height that the rafter sits above the wall. This is a plumb line, measured from the seat cut up to the top edge of the rafter. Also called the stand. See Figures 12-1, 12-18 and 12-28.

T.J.I. - A fabricated floor system. Each joist is made from a laminated top and bottom chord, pressed onto a piece of particleboard or waferboard.

Toenail - A nail sent into a board at about a 60-degree angle, an inch or so from its edge. This method is used when driving nails near the base of a framing member, and face nailing (sending the nail straight into the wood) isn't possible. The nail should penetrate so that half the nail is in each member. See Figure 7-18.

Tongue and groove (T&G) - Material with one edge shaped with a narrow projection along the centerline of its length that is designed to fit into a corresponding groove on another piece, creating an interlocking joint. Paneling or flooring members are often designed with one tongue edge and one groove edge so that they can be installed in a continuous interlocking pattern.

Top cripple - Short wood members that run vertically between window or door headers and the first top plate, typically spaced on a 16-inch on-center layout.

Top plate - The upper horizontal structural member in a framed wall. The studs are face nailed through the first top plate when the wall is on the ground. The second, or double top plate, is added after the studs are all nailed to overlap onto an adjoining wall's top plate, locking the walls together. See Figures 12-3 and 12-5.

Topping off - The act of tying together all the double top plate laps after all the wall groups are standing. This is done by walking the top plates of the walls and sending two nails into each lap after pulling the connection tight using a toenail. See Figure 5-14.

Tract - A housing project with several houses built on one tract of land by the same builder. There are commonly three to four floor plans that are reversed to create an illusion of more variety than there actually is. In a healthy economy, tract construction can provide a steady income that custom housing usually can't match. See Figure 1-2.

Tread - The horizontal boards that form the steps on a staircase. It is also another name for the run on a set of stairs. An individual tread is also sometimes called a run. See Figure 9-1.

Trimmer - The stud that is nailed to the king stud and runs up under the header when you're framing rough openings. Once both trimmers are installed, the window or door opening is complete. See Figure 5-5.

Truss - One component of a prefabricated roof system, usually built off the job site and delivered in packages. Each truss consists of two opposing rafters attached to a lower chord. The truss structure is designed to act as a beam. See Figure 11-1.

U

Understacking - Building small sections of rake or flat walls in the opening formed between the tops of normal studded walls and the bottom chords of scissor trusses or rafters. Some walls are nearly impossible to build accurately as rakes, so it's easier to build them with normal studs and have the roof stackers fill in the small angled sections once the rafters or trusses are up. See Figure 11-16.

V

Valley - The point where two roof slopes intersect and form a trough for water to drain. See Figure 12-4.

Vaulted ceiling - A high finished ceiling that follows the bottom of the rafters. Instead of installing ceiling joists, the roof is stacked and the rafters are left exposed, or drywall is attached directly to them.

Vent block - Blocks, with holes drilled in them, used intermittently in place of the flush or frieze blocks placed between rafters to let air circulate in the attic. The holes are screened on the inside to keep birds out of the attic. See Figure 11-5.

Index

Other Practical References

• Carpentry Estimating

Simple, clear instructions on how to take off quantities and figure costs for all rough and finish carpentry. Shows how to convert piece prices to MBF prices or linear foot prices, use the extensive manhour tables included to quickly estimate labor costs, and how much overhead and profit to add. All carpentry is covered; floor joists, exterior and interior walls and finishes, ceiling joists and rafters, stairs, trim, windows, doors, and much more. Includes sample forms, checklists, and the author's factor worksheets. **320 pages, 8½ x 11, $32.50**

• National Repair & Remodeling Estimator

The complete pricing guide for dwelling reconstruction costs.

Reliable, specific data you can apply on every repair and remodeling job. Up-to-date material costs and labor figures based on thousands of jobs across the country. Provides recommended crew sizes; average production rates; exact material, equipment, and labor costs; a total unit cost and a total price including overhead and profit. Separate listings for high- and low-volume builders, so prices shown are accurate for any size business. Estimating tips specific to repair and remodeling work to make your bids complete, realistic, and profitable. *Repair & Remodeling Estimate Writer FREE on a 5¼" high-density (1.2 Mb) disk when you buy the book.* (Add $10 for *Repair & Remodeling Estimate Writer* on extra 5¼" double density 360K disks or 3½" 720K disks.) **368 pages, 11 x 8½, $29.50. Revised annually.**

• Finish Carpenter's Manual

Everything you need to know to be a finish carpenter: the proper use of hand and power tools, assessing a job before you begin, and tricks of the trade from a master finish carpenter. Easy-to-follow instructions for installing doors and windows, ceiling treatments (including fancy beams, corbels, cornices and moldings), wall treatments (including wainscoting and sheet paneling), and the finishing touches of chair, picture, and plate rails. Specialized interior work includes cabinetry and built-ins, stair finish work, and closets. Also covers exterior trims and porches. Includes manhour tables for finish work, and hundreds of illustrations and photos. **208 pages, 8½ x 11, $22.50**

• Plumber's Handbook Revised

This new edition shows what will and won't pass inspection in drainage, vent, and waste piping, septic tanks, water supply, fire protection, and gas piping systems. All tables, standards, and specifications completely up-to-date with recent plumbing code changes. Covers common layouts for residential work, how to size piping, selecting and hanging fixtures, practical recommendations, and trade tips. The approved reference for the plumbing contractor's exam in many states. **240 pages, 8½ x 11, $18.00**

• Roof Framing

Shows how to frame any type of roof in common use today, even if you've never framed a roof before. Includes using a pocket calculator to figure any common, hip, valley, or jack rafter length in seconds. Over 400 illustrations cover every measurement and every cut on each type of roof: gable, hip, Dutch, Tudor, gambrel, shed, gazebo, and more. **480 pages, 5½ x 8½, $22.00**

• Manual of Professional Remodeling

The practical manual of professional remodeling that shows how to evaluate a job so you avoid 30-minute jobs that take all day, what to fix and what to leave alone, and what to watch for in dealing with subcontractors. Includes how to calculate space requirements; repair structural defects; remodel kitchens, baths, walls, ceilings, doors, windows, floors and roofs; install fireplaces and chimneys (including built-ins), skylights, and exterior siding. Includes blank forms, checklists, sample contracts, and proposals you can copy and use. **400 pages, 8½ x 11, $23.75**

• Spec Builder's Guide

Shows how to plan and build a home, control construction costs, and sell to get a decent return on the time and money you've invested. Includes professional tips to ensure success as a spec builder: how government statistics help you judge the housing market, cutting costs at every opportunity without sacrificing quality, and taking advantage of construction cycles. Includes checklists, diagrams, charts, figures, and estimating tables. **448 pages, 8½ x 11, $27.00**

• National Construction Estimator

Current building costs for residential, commercial, and industrial construction. Estimated prices for every common building material. Manhours, recommended crew, and labor cost for installation. Includes Estimate Writer, an electronic version of the book on computer disk, with a stand-alone estimating program ---- *FREE* on 5¼" high density (1.2Mb) disk. The National Construction Estimator and Estimate Writer on 1.2Mb disk cost $31.50. (Add $10 if you want Estimate Writer on 5¼" double density 360K disks or 3½" 720K disks.) **592 pages, 8½ x 11, $31.50. Revised annually**

• Remodeler's Handbook

The complete manual of home improvement contracting: evaluating and planning the job, estimating, doing the work, running your company, and making profits. Pages of sample forms, contracts, documents, clear illustrations, and examples. Includes rehabilitation, kitchens, bathrooms, adding living area, reflooring, residing, reroofing, replacing windows and doors, installing new wall and ceiling cover, repainting, upgrading insulation, combating moisture damage, selling your services, and bookkeeping for remodelers. **416 pages, 8½ x 11, $27.00**

• Video: Roof Framing 1

A complete step-by-step training video on the basics of roof cutting by Marshall Gross, the author of the book *Roof Framing*. Shows how to calculate rise, run, and pitch, and lay out and cut common rafters. **90 minutes, VHS, $80.00**

• Video: Roof Framing 2

A complete training video on the more advanced techniques of roof framing by Marshall Gross, the author of *Roof Framing*. Shows how to lay out and frame an irregular roof, and make tie-ins to an existing roof. **90 minutes, VHS, $80.00**

• Building Cost Manual

Square foot costs for residential, commercial, industrial, and farm buildings. Quickly work up a reliable budget estimate based on actual materials and design features, area, shape, wall height, number of floors, and support requirements. Includes all the important variables that can make any building unique from a cost standpoint. **240 pages, 8½ x 11, $16.50. Revised annually**

• Contractor's Survival Manual

How to survive hard times and succeed during the up cycles. Shows what to do when the bills can't be paid, finding money and buying time, transferring debt, and all the alternatives to bankruptcy. Explains how to build profits, avoid problems in zoning and permits, taxes, time-keeping, and payroll. Unconventional advice on how to invest in inflation, get high appraisals, trade and postpone income, and stay hip-deep in profitable work. **160 pages, 8½ x 11, $16.75**

• Estimating Home Building Costs

Estimate every phase of residential construction from site costs to the profit margin you include in your bid. Shows how to keep track of manhours and make accurate labor cost estimates for footings, foundations, framing and sheathing finishes, electrical, plumbing, and more. Provides and explains sample cost estimate worksheets with complete instructions for each job phase. **320 pages, 5½ x 8½, $17.00**

• Rafter Length Manual

Complete rafter length tables and the "how to" of roof framing. Shows how to use the tables to find the actual length of common, hip, valley, and jack rafters. Explains how to measure, mark, cut and erect the rafters; find the drop of the hip; shorten jack rafters; mark the ridge and much more. Loaded with explanations and illustrations. **369 pages, 5½ x 8½, $15.75**

• Handbook of Construction Contracting, Vol.1

Everything you need to know to start and run your construction business; the pros and cons of each type of contracting, the records you'll need to keep, and how to read and understand house plans and specs so you find any problems before the actual work begins. All aspects of construction are covered in detail, including all-weather wood foundations, practical math for the job site, and elementary surveying. **416 pages, 8½ x 11, $24.75**

• Hanbook of Construction Contracting Vol. 2

Everything you need to know to keep your construction business profitable; different methods of estimating, keeping and controlling costs, estimating excavation, concrete, masonry, rough carpentry, roof covering, insulation, doors and windows, exteriof finishes, specialty finishes, scheduling work flow, managing workers, advertising and sales, spec building and land development, and selecting the best legal structure for your business. **320 pages, 8½ x 11, $26.75**

• Building Layout

Shows how to use a transit to locate a building correctly on the lot, plan proper grades with minimum excavation, find utility lines and easements, establish correct elevations, lay out accurate foundations, and set correct floor heights. Explains how to plan sewer connections, level a foundation that's out of level, use a story pole and batterboards, work on steep sites, and minimize excavation costs. **240 pages, 5½ x 8½, $15.00**

• Construction Estimating Reference Data

Provides the 300 most useful manhour tables for practically every item of construction. Labor requirements are listed for sitework, concrete work, masonry, steel, carpentry, thermal and moisture protection, door and windows, finishes, mechanical and electrical. Each section details the work being estimated and gives appropriate crew size and equipment needed. This new edition contains *DataEst*, a computer estimating program on a high density disk. This fast, powerful program and complete instructions are yours free when you buy the book. **432 pages, 8½ x 11, $39.50**

• Contractor's Guide to the Building Code Rev

This completely revised edition explains in plain English exactly what the Uniform Building Code requires. Based on the most recent code, it covers many changes made since then. Also covers the Uniform Mechanical Code and the Uniform Plumbing Code. Shows how to design and construct residential and light commercial buildings that'll pass inspection the first time. Suggests how to work with an inspector to minimize construction costs, what common building shortcuts are likely to be cited, and where exceptions are granted. **544 pages, 5½ x 8½, $28.00**

• Stair Builders Handbook

If you know the floor-to-floor rise, this handbook gives you everything else: number and dimension of treads and risers, total run, correct well hole opening, angle of incline, and quantity of materials and settings for your framing square for over 3,500 code-approved rise and run combinations - several for every 1/8-inch internal from a 3 foot to a 12 foot floor-to-floor rise. **416 pages, 5½ x 8½, $15.50**

• HVAC Contracting

Your guide to setting up and running a successful HVAC contracting company. Shows how to plan and design all types of systems for maximum efficiency and lowest cost ---- and explains how to sell your customers on your designs. Describes the right way to use all the essential instruments, equipment, and reference materials. Includes a full chapter on estimating, bidding, and contract procedure. **256 pages, 8½ x 11, $24.50**

• Estimating Plumbing Costs

Offers a basic procedure for estimating materials, labor, and direct and indirect costs for residential and commercial plumbing jobs. Explains how to read and understand plot plans, design drainage, waste, and vent systems, meet code requirements, and make an accurate take-off for materials and labor. Includes sample cost sheets, manhour production tables, complete illustrations, and all the practical information you need. **224 pages, 8½ x 11, $22.50**

• Estimating Electrical Construction

Like taking a class in how to estimate materials and labor for residential and commercial electrical construction. Written by an A.S.P.E. National Estimator of the Year, it teaches you how to use labor units, the plan take-off, and the bid summary to make an accurate estimate, how to deal with suppliers, use pricing sheets, and modify labor units. Provides extensive labor unit tables and blank forms for your next electrical job. **272 pages, 8½ x 11, $19.00**

• Carpentry Layout

Explains the easy way to figure: cuts for stair carriages, treads and risers; lengths for common, hip, and jack rafters; spacing for joists, studs, rafters, and pickets; layout for rake and bearing walls. Shows how to set foundation corner stakes, even for a complex home on a hillside. Practical examples on how to use a hand-held calculator as a powerful layout tool. **240 pages, 5½ x 8½, $16.25**

• Blueprint Reading for the Building Trades

How to read and understand construction documents, blueprints, and schedules. Includes layouts of structural, mechanical, HVAC and electrical drawings. Shows how to interpret sectional views, follow diagrams and schematics, and covers common problems with construction specifications. **192 pages, 5½ x 8½, $11.25**

• Roofers Handbook

The journeyman roofer's complete guide to wood and asphalt shingle application on new construction and reroofing jobs: how to make smooth tie-ins on any job, cover valleys and ridges, handle and prevent leaks. Includes how to set up and run your own roofing business and sell your services. Over 250 illustrations and hundreds of trade tips. **192 pages, 8½ x 11, $19.00**

• Paint Contractor's Manual

How to start and run a profitable paint contracting company: getting set up and organized to handle volume work, avoiding mistakes, squeezing top production from your crews and the most value from your advertising dollar. Shows how to estimate all prep and painting. Loaded with manhour estimates, sample forms, contracts, charts, tables and examples you can use. **224 pages, 8½ x 11, $19.25**

• Estimating Tables for Home Building

Produce accurate estimates for nearly any residence in just minutes. This handy manual has tables you need to find the quantity of materials and labor for most residential construction. Includes overhead and profit, how to develop unit costs for labor and materials, and how to be sure you've considered every cost in the job. **336 pages, 8½ x 11, $21.50**

• Video: Drywall Contracting 1

How to measure, cut, and hang: the tools you need and how to use them to do top quality work on any job in the shortest time possible. Explains how to plan the job for top productivity, straighten studs, use nails, screws or adhesive to best advantage, and make the most of labor-saving tools. **33 minutes, VHS, $24.75**

• Video: Drywall Contracting 2

How to use mechanical taping tools, mix and apply compound, use corner bead, finish and texture board, and solve the most common drywall problems. Includes tips for making a good living in the drywall business. **38 minutes, VHS, $24.75**

• Video: Stair Framing

Shows how to use a calculator to figure the rise and run of each step, the height of each riser, the number of treads, and the tread depths. Then watch how to take these measurements to construct an actual set of stairs. You'll see how to mark and cut your carriages, treads, and risers, and install a stairway that fits your calculations for the perfect set of stairs. **60 minutes, VHS, $24.75**

• How to Succeed With Your Own Const. Business

Everything you need to start your own construction business: setting up the paperwork, finding the work, advertising, using contracts, dealing with lenders, estimating, scheduling, finding and keeping good employees, keeping the books, and coping with success. If you're considering starting your own construction business, all the knowledge, tips, and blank forms you need are here. **336 pages, 8½ x 11, $19.50**

• Profits in Buying & Renovating Homes

 Step-by-step instructions for selecting, repairing, improving, and selling highly profitable "fixer-uppers." Shows which price ranges offer the highest profit-to-investment ratios, which neighborhoods offer the best return, practical directions for repairs, and tips on dealing with buyers, sellers, and real estate agents. Shows you how to determine your profit before you buy, what "bargains" to avoid, and how to make simple, profitable, inexpensive upgrades. **304 pages, 8½ x 11, $19.75**

• Illust. Guide to the 1993 National Electrical Code

This fully-illustrated guide offers a quick and easy visual reference for installing electrical systems. Whether you're installing a new system or repairing an old one, you'll appreciate the simple explanations written by a code expert, and the detailed, intricately-drawn and labeled diagrams. A real time-saver when it comes to deciphering the current NEC. **256 pages, 8½ x 11, $26.75**

• Carpentry for Residential Construction

How to do professional quality carpentry work in residences. Illustrated instructions on everything from setting batterboards to framing floors and walls, installing floor, wall, and roof sheathing, and applying roofing. Covers finish carpentry: installing each type of cornice, frieze, lookout, ledger, fascia, and soffit; hanging windows and doors; installing siding, drywall, and trim. Each job description includes tools and materials needed, estimated manhours required, and a step-by-step guide to each part of the task. **400 pages, 5½ x 8½, $19.75**

• Carpentry In Commercial Construction

Covers forming, framing, exteriors, interior finish, and cabinet installation in commercial buildings: how to design and build concrete forms, select lumber dimensions, what grades and species to use for a design load, how to select and install materials based on their fire rating or sound-transmission characteristics, and plan and organize a job efficiently. Loaded with illustrations, tables, charts, and diagrams. **272 pages, 5½ x 8½, $19.00**

• Planning Drain, Waste & Vent Systems

How to design plumbing systems in residential, commercial, and industrial buildings. Covers designing systems that meet code requirements for homes, commercial buildings, private sewage disposal systems, and even mobile home parks. Includes relevant code sections and many illustrations to guide you through what the code requires in designing drainage, waste, and vent systems. **192 pages, 8½ x 11, $19.25**

• Estimating Framing Quantities

Gives you hundreds of time-saving estimating tips. Shows how to make thorough step-by-step estimates of all rough carpentry in residential and light commercial construction: ceilings, walls, floors, and roofs. Lots of illustrations showing lumber requirements, nail quantities, and practical estimating procedures. **285 pages, 5½ x 8½, $34.95**

• Estimating & Bidding for Builders & Remodelers

New and more profitable ways to estimate and bid any type of construction. This award winning book shows how to take off labor and material, select the most profitable jobs for your company, estimate with a computer (**FREE** estimating disk enclosed), fine-tune your markup, and learn from your competition. Includes a sample disk of Estimate Writer, an estimating program with a 30,000-item database on a 5¼" high-density disk when you buy the book. (If your computer can't use high-density disks, add $10 for 5¼" or 3½" double-density disks.) **272 pages, 8½ x 11, $29.75**

• Estimating Painting Costs

Here's an accurate step-by-step estimating system, based on easy-to-use manhour tables, for estimating painting costs from simple residential repaints to complicated commercial jobs ---- even heavy industrial and government work. Explains taking field measurements, doing take-offs from plans and specs, predicting productivity, figuring labor and material costs, and overhead and profit. Includes manhour and material tables, plus sample forms and checklists. **448 pages, 8½ x 11, $28.00**